The
IMPACTS
Dynamic

Working Against Dispersal in Human Society and Across the Universe

Eddie + Janeth —

To the Future!

[signature]

The
IMPACTS
Dynamic

Working Against Dispersal in Human
Society and Across the Universe

Dickey Eason

THE IMPACTS DYNAMIC
Working Against Dispersal in Human
Society and Across the Universe

Copyright 2009 by Dickey Eason

All rights reserved.

ISBN: 978-0-9746284-2-4

Additional information about IMPACTS
can be found at http://www.theimpacts.com.

IMPACTS Publishing
Cary, North Carolina

To my Mother,
filled with **IMPACTS**
love and energy
but stolen away from life
far too early.

Table of Contents

Acknowledgements

I would like to acknowledge the contributions of others to this work.

A thank you to all of the editors who have assisted me on this project: Rachel Trusheim who formulated a very important outline in the early going; Ame Kitchener who helped with some developmental editing; Dwen Andrews-Cita and Pam Langley who assisted in line-editing; Angela T. Pace who came on board at the end and did a superlative job in all aspects; and someone who wants to remain anonymous but whose contributions in the developmental editing area were extremely important.

Thanks to Clare Strayhorn of Piedmont Litho in Raleigh, N.C. Clare has been a mainstay in all of my projects over the past few years, always producing excellence. All of the diagrams and illustrations prepared for the book were done by Clare, along with the layout of the cover.

Thanks to Dr. Frank Wilczek, a Nobel Prize winner in physics, who was kind enough to answer several physics questions.

Thanks to Henry Hutton who helped with the publishing process, and to Elmore Hammes who did the final editing and layout.

A special thank you to Carolyn Anderson and Stephanie Allen, friends for the past ten years who have always been enthusiastically supportive. They don't come any better than those two.

Great appreciation to my business customers who helped me in ways that none of us could ever have imagined.

And thanks to my friend PG who listened to all of my ideas and to all of my frustrations.

Introduction

The highly respected author and biologist, Edward O. Wilson, in a recent interview with *The Wall Street Journal*, mused that a new basis for spiritual values might be found, not in the usual religious sources, but in what he sees as the inspiring story of human origins and history.

"We need to create a new epic based on the origins of humanity," he said, adding: "Homo sapiens have had one hell of a history! And I am speaking both of deep history - evolutionary, genetic history - and then, added on to that and interacting with it, the cultural history recorded for the past 10,000 years or so."

That is what we are going to do—we are going to take a completely new look at how the world works, and that will include human civilization and its development. In the process, we will discover a dynamic that has existed among and within modern human populations since their emergence over 100,000 years ago. We will also see that the same dynamic operates in the physical world. We are going to marry the two worlds, human civilization and the physical universe, including the atom, and what we see will probably surprise you—they are behaving in exactly the same manner. If you understand one, you understand the other. And there is something else you will see as well. Not only could this discussion potentially reorient your view of the world and the universe, it is also going to be fun and exciting.

Human beings have been asking these kinds of questions for thousands of years. It appears that our ancestors living prior to the emergence of agriculture were more capable in many areas than we are today, including in the practical and cosmological realms. But then everything they contemplated was done with an eye toward everyday application. Nobody has yet figured out how humans around the globe transported stone slabs weighing several tons, sometimes hundreds of miles and often up steep mountains. Many of these pieces were cut and fit so precisely that even today, thousands of years later, you cannot slide a razor blade between the joints. We couldn't move that kind of tonnage today with a crane. How did they do it? We have no idea. If they could do it, why can't we? Obviously something has been lost due to 'civilizing' influences. What else have we lost? Quite a lot actually, including much of our original humanity. But that is what happens universally to a developing structure—the process of growth dilutes its foundational creative-formative-productive energy. It is not unique to modern humans.

I totally reject the idea that we cannot figure out the secrets of the universe—that we cannot comprehend how to 'transport huge pieces of stone up the mountainside'. I also reject the notion that we cannot change the world with our newfound knowledge. We have succumbed to the forces in humanity which do not want us to ask, "WHY?" and which do not want us to truly believe that we can find the answers to the world's problems. But we will discover who and what these forces are and why they do not want us to correctly diagnose the situation. We will find that there is no reason to throw our hands up in despair, though the solutions will not be easy.

Basically we are going to discover one very important variable that has been overlooked in the study of modern human beings, and the discovery of that variable is going to force us to look at EVERYTHING differently: recorded history and pre-history, science, religion, war and peace, men and women, human society, business and trade, the arts, medicine, the universe, and everything else. In the epilogue to his book, *Guns, Germs, and Steel,* author and scientist Jared Diamond writes, "The challenge now is to develop human history as a science, on a par with acknowledged historical sciences such as astronomy, geology, and evolutionary biology." That is precisely what we will do, and as we proceed, we will be able to see how this new knowledge will aid us in understanding the dynamics operating at any moment in history and in our contemporary world.

Discovering a dynamic at work is energizing. Not seeing the dynamic that is operating costs us in incalculable ways, depriving us of potential motivation, further discovery, and creative

12

production. If we can break down some initial barriers, we can push over still more and more.

We need a new, more realistic way of looking at human civilization and human society—more objective and less sentimental. But here is the paradox—if we approach our study objectively, we are going to find that the whole experience actually produces more sentiment. The truth usually does. Our efforts to 'figure out everything' with the current paradigm are doomed just as pre-Copernican-Europe's efforts were doomed when it tried to be 'scientific'. You will recall that Copernicus had 'reminded' everyone that the sun was the center of the solar system—not the earth as the Catholic Church had decreed. I say reminded because human beings had known thousands of years before that the earth revolved around the sun. But like much of the knowledge from early humans, it was lost and would have to be rediscovered. Actually, it wasn't lost as much as it was rejected because the truth did not fit the needs of those in charge. As we will see, that has been a recurring theme since the emergence of agriculture and settled living conditions.

While we have at least made some progress discerning the cosmic architecture, our study of human beings and their behavior is still seriously lacking. If we do not understand the beginnings and development of our modern human foundation, we have no chance of understanding the present.

I think you will discover as we go forward that our current version of human civilization and its development, including the contemporary world, is so distorted that it reads more like a novel than nonfiction. Presently, as opposed to early modern humans, we have a very disjointed view of reality. Theirs was more connected, and their world was more harmonious as well, with each other and with nature. Let's determine why we have drifted apart since those early times other than the obvious reasons of increased population. Is something else at work?

I think we have been asking the wrong questions and looking for the answers in the wrong places. We have been looking within the current paradigm when experience should show us that the answers come from the outside, or periphery, and always have. But we have grown up with this paradigm; we do not realize that there are far more plausible truths currently residing just beyond our reach. The energy field has us solidly within its grasp. Because we have the wrong paradigm, we are identifying problems incorrectly, and consequently we are searching for the wrong solutions. We are very much like the dog chasing its own tail. We are going 'round and 'round, not realizing the absurdity and futility of our actions.

We will discover that there are two major forms of energy within human civilization, and they are the same two energies that struggle with one another throughout the universe. The simple hydrogen atom, with its one proton and one electron, epitomizes this struggle. What we are really talking about is the 2nd Law of Thermodynamics, but do not let that scare you. The 2nd Law is very simple to understand: localized or concentrated energy of any kind has a tendency to spontaneously disperse if not hindered from doing so.

So we have a powerful dispersal force, but we also appear to have a force that works hard to hinder the 2nd Law from doing its job. The battle in the universe and on planet earth among people is between the dispersal forces which obey the 2nd Law and the 2nd Law-breakers, which try to keep everything together. The law-breakers, whether humans or matter, form powerful bonds of energy which resist the tendencies to disperse. Dispersal energy is male—the energy that hinders dispersal is female. Among people the lines get blurred, but the general outline holds true.

Our present view of the world actually comes from the dispersal forces instead of the problem-solvers, who happen to be the bonders and the violators of the 2nd Law. The dispersal forces control the paradigm. We are going to look at the picture with both forces highlighted, and a very different view will emerge.

With this study we are going to demystify human civilization and its origins. My aim is to try to see, understand, and explain the world and universe as they are, not how some want them to be, believe they are, or want us to believe they are. If we turn the camera angle just a little, drop our biases, and be and remain open to whatever we see, we will encounter an entirely new picture, one that makes a whole lot more sense. We cannot go about fixing the world until we know what is actually wrong with it, and what is wrong with it is not what we have been told.

Many will think this explanation is too 'far out' to be seriously considered, but actually what we have now is the fanciful paradigm, one in which we think we behave according to certain principles and the rest of the universe to others. We look at the night sky as if it were a light show rather than the dynamic universe that it is, and then we retreat into our cocoon and 'invent' explanations for our behavior, never considering that the same dynamic exists within us as in a distant galaxy. We are too blindly adhering to the script that has been written for us. That script needs to be revised or rewritten. As I said, it is closer to fiction than nonfiction.

I cannot find the author of the following quotation so I extend my apologies to him or her, but it is too valuable to be left unsaid: "The assumptions that are the most pernicious are the ones we don't know we are making because they appear so intrinsically obvious." And I might add, "Or because we have been 'taught' so well."

The failure to examine our assumptions helps support the status quo. Soft-pedaling or tiptoeing through the issues serves no purpose whatsoever. Examining these issues head-on is a necessity. We have become a nation and world of enablers as we turn away from cruelty and exploitation committed in 'our name'. We are part of the conspiracy to 'not know'. But all of this can be explained through an understanding of the dynamics at work in human society and civilization. The power-control core, which is part of the dispersal energy, uses fear as a tool of its survival, and that fear partially drives our behavior. Today's hierarchal structures push constituent members apart, frequently 'naturally selecting' those who are more concerned with self rather than the 'other'. If we start to gain an awareness of the basic dynamics, we can begin the work of rectifying the imbalances.

Humanity's development is much more formulaic than we would ever suspect. We cannot see it because we have tried to paint the picture ourselves instead of seeing it for what it really is. Just as it was during pre-Copernicus days, we are still putting ourselves at the center of the universe.

Today's paradigm mostly blames human beings for their difficulties and struggles in life while absolving those at the top of their share of the responsibility. The Roman Catholic Church of the Middle Ages has been replaced by another 'Church' and this one is just as powerful, or even more so. The modern world places people very much in a double-bind situation: "You are free to realize your potential and be who you want to be." But of course that is not true. The truth is that you are free to realize your potential if your thoughts and behavior are in accord with those of the 'powers-that-be', and if you have adequate resources and support available to you. You are much more likely to get the needed support if your attitude and actions are in tune with the status quo forces.

Copernicus's model was revolutionary only because the existing model was so distorted. It is the same with what I am saying—it just looks strange because our current model is so skewed. The intent is to show that everything is following the same model—us, the universe, bees, elephants, the Milky Way galaxy, the hydrogen atom—all of it. My goal is to find the truth and connecting threads, as much as is possible at this point with our faulty measuring devices, one of these being our brain. Our brain is a wonderful instrument, and we will see that it is an encapsulation of the two major forces of the universe, but it is relatively new and is still under development.

We limit ourselves terribly by the teachings we receive. Those teachings are generally what the prevailing authority structure wants us to believe. Almost everyone 'buys into it' because it is so powerful. There appear to be no alternatives. Our present-day world has room for only one paradigm, which is essentially a 'prescription' for living. But we can look at any part of history and see that the view that was 'in control' always had holes in it. Reality has always been what we said it was—what was accepted at the time. But times change, and so do our ways of looking at things, though we generally have to be pulled kicking and screaming into a new view, even if it is the correct view. Old habits die hard.

Occasionally someone comes along and turns everything upside down. Isaac Newton was only twenty-three years old when he developed his three laws of motion. Einstein in his formula $E = mc^2$, where energy equals the mass of an object times the speed of light squared, revealed that mass and energy are the same thing. These two men helped explain the workings of the universe with nice, clear laws and formulas. And while human behavior does not lend itself currently to that kind of precise predictability, we are following the same general laws and processes. How could we not be?

Let's define what we mean by a dynamic. I think it is useful to look at its use as an adjective and a noun in order to truly grasp its importance.

From the *American Heritage Dictionary*:

Adjective:
- Of or relating to energy or to objects in motion
- Characterized by continuous change, activity, or progress
- Marked by intensity and vigor; forceful.

Noun:
- An interactive system or process, especially one involving competing or conflicting forces
- A force, especially political, social, or psychological.

From *Collins Essential English Dictionary*:

Adjective:
- Describing a person—full of energy, ambition, or new ideas
- Relating to a force of society, history, or the mind that produces a change
- Physics—relating to energy or forces that produce motion.

From *Encarta*—MSN:

Adjective:
- Full of energy, enthusiasm, and a sense of purpose and able to get things going and to get things done
- Characterized by vigorous activity and producing or undergoing change and development.

Sorry for the 'dynamic' overload, but I want you to get a solid grasp of what the word means because it is integral to everything we discuss. You can see from above that a dynamic is about energy, motion, intensity, change, and competition with another force. That will be the essence of our discussion.

We currently see more of a dynamic operating in a basketball game than we do in the universe. Isn't that odd? Why is that the case, and has it always been the case? No, it has not. The hierarchal structure of today's world does not allow for a dynamic to be manifested in the physical and human worlds because the energy flow would be upset, as we will see in our discussions to follow. So humanity's current structure does not favor the identification of a dynamic though one may clearly be operating. A dispersal force does not want to see its opposite, nor does it want others to see its opposite.

We try to look objectively at other species but not our own. Why is that? Is it because we might see the reality of what is really happening in the world? What then would we do? Hence, we tiptoe around the real issues and continue with our games of charades and make-believe, in the process turning our heads away from the ugliness that we in part enable. We convince ourselves that being 'positive' is more important than being 'just'. So we let the 'bad stuff slide' and deny our own culpability. The dispersal forces are in control of our lives.

Why I Started This Study

What prompted my study of all of this along with the subsequent discoveries? It started in a time of turbulence for me, which I found is the way that many important discoveries are made. Several years ago, I started a new business upon the demise of another one, nothing exceptional for small business entrepreneurs. I was excited about it and was even looking forward to the long, hard grind that usually accompanies such ventures. I figured I would get very involved in charitable and other community activities as I built the business, and would then develop franchising for it. I did get involved in the community, but the franchising got sidetracked. Why? Because I started making extraordinary discoveries, or rather a set of discoveries, that revolutionized my thinking about how the world works and provided me with an entirely new worldview. Actually, the findings were an accident, as is often the case. I was just trying to deliver results expeditiously and maximize the potential of my new business. I would learn later that these two related concepts would be the keystone to my subsequent work.

The business was based on a simple concept—the need for household storage, specifically in the homeowner's garage. At first I offered cabinets, pegboard, workbenches, tool hangers, and a three-foot-deep Big Shelf, 7-8 feet off the floor. Soon it became apparent that 98% of my sales were for the Big Shelf. So I dropped everything else, and the business became Big Shelf – Garage Storage Solutions. It was a very simple concept with little overhead. The major home improvement store nearby became my warehouse. Business owners know the preferred drill—keep it as simple as possible. It makes life a whole lot easier and frees time for other pursuits.

Of course I had no idea how important the issue of storage is to understanding the development of human civilization, and in fact, life itself. The bonds against dispersal that we mentioned earlier? That is the storage of energy. You might think the connection to garage storage is implausible, but it is actually quite real as we will see.

Many people were openly skeptical and others quietly so: "You can't make a business out of a big shelf." But not only did I make a business out of it, I also made some incredible discoveries about business and marketing AND human civilization and the physical universe. There is a strong lesson here—do not let others tell you what you can and cannot accomplish. They are projecting their thoughts and feelings onto your world, and your world and theirs may have very little in common. People can make major discoveries digging a ditch, and have done so. Quite simply, most people are under the influence of the prevailing energy field, and they are adhering to its philosophy and dictates.

Let me tell you a little about the process that enabled these discoveries. After four years of doggedly pursuing the 'best' business strategies and obtaining only mediocre results, I decided that traditional marketing approaches were not working and were probably not going to work. I needed a way to reliably identify my customer. You would think that the customer for my product would be anyone with a storage problem, a garage, and a few hundred bucks to spare. But it was not that simple. Nothing ever is. Sometimes, it is better to tear up the plan and start over rather than continually trying to force a square peg into a round hole. So that is what I did. I invented my own process of customer identification since I could not find another one that seemed to be applicable.

I started paying attention to everything in my customers' environments. I had no way of knowing that by taking this approach, I was replicating the attitude and behavior of the personality profile that was the centerpiece of the development of modern humans from over 100,000 years ago.

I paid no attention to demographics because most of my customers came from the same middle socioeconomic group. So I looked for other clues. I started noticing that half of my customers lived on cul-de-sacs and about 15% on corners. Most were living close to a buffer of some sort: a stand of trees in the back, a vacant lot next door, a small park-like area in front, on a dead-end street—anything that provided a little extra privacy. And very often water was nearby—a stream, river, lake, pond, even an outdoor fish pool. Frequently, the scenery viewed from the house could be quite picturesque.

With all of these remarkable commonalities among customers, I began to believe that I was dealing with a particular personality profile, a 'getting-things-in-order' profile. Upon talking with friends in other businesses and industries, it became clear to me that what I had identified was a specific group that made very specific purchases based around strong values, the strongest being the home and family. I called this group of people the IMPACTS.

Then, as I read books about people who were influential in business, cutting-edge technology, science, medicine, the arts, sports, and many other areas, I began to see that they shared the same profile as those I was calling IMPACTS. In *Good to Great* and *Built to Last*, books in which the author, Jim Collins, expounded on the reasons certain businesses succeeded above and beyond others, I saw the strong imprint of the IMPACTS leading the way. In the many works of James Burke (*Connections, The Day the Universe Changed, The Knowledge Web,* and others), I saw the IMPACTS on every page of every chapter. I was excited and enthralled with what I was seeing. What I had believed was a purchasing group appeared to be a unique personality profile that had always led human beings forward. But when and how did the profile develop? I was hooked on finding the answers.

Why I Wrote This Book

You might ask why I embarked on such an ambitious venture requiring years of research, reflection, and writing. Well, just as you probably have your own questions about the world and the universe, so too, do I. I just may be a little more aggressively obsessive about finding the answers. I never stop digging until I am satisfied that I have gone as far as I can go.

These are some of the questions that intrigue me: Are people today different from human beings of five thousand years ago? Ten thousand years ago? Fifty thousand? Why did it take humans so long to invent the wheel? It is only about five thousand years old while modern humans are over one hundred thousand years old. How did we go from cave paintings twenty thousand years ago to landing on the moon? How is it possible that people were able to construct such impressive structures as Stonehenge in England, the Pyramids in Egypt, and the Parthenon in Greece, with such rudimentary tools? But then, maybe they weren't so rudimentary after all. Why do art and architecture from long ago seem so much more exquisite than that of today? Why has the world taken off like a rocket in technology areas? Why do some people seem to do many things so well? Why do some seem to see needs and act on them while others do not? How could people like the Vikings be so brutal and yet so advanced at the same time? Why do we have constant warfare around the globe? Has it always been the same? Why do the 'bad guys' seem to end up running the world instead of the 'good guys'? Why do the great inventors and discoverers like Einstein and Da Vinci usually remain adamantly opposed to the 'insanity' of war even as their inventions and discoveries are often used for horrible acts of destruction? Are we going backwards as far as being civilized is concerned? Is there any hope for a better world?

I have come to realize during the course of my life that I cannot accept the explanations of the way things are in almost any area. Frankly, on many occasions I have wondered if anyone knew anything about anything, or if everyone were just repeating what they had heard. Independent thinkers have been hard to find. I no longer accept anyone's opinion about anything. I just dig for the answers myself because I have learned that I will keep searching long after almost everyone else has stopped.

As a young child, I subconsciously assumed that adults knew what they were talking about, and that their answers were based on reason and reflection. I do not remember at what point I realized that I was very wrong in my assumption. I also do not remember if I were one of those kids who always ask, "WHY?" But now I know why young people stop asking. The adults are not asking "WHY?" so the young quickly learn that this is not a WHY or cause-and-effect world. It is a world of production, uniformity, subservience to authority, and denial of realities that do not fit the existing paradigm.

It turns out that the book I have written is the book I always wanted to read, a book that connected, as much as is possible at this particular time, all of the variables of our existence: the earth and life on it, the universe beyond, people, science, art, history, business, everything. I want to share with the world what I have found in the hope that it may help people find answers to their questions. IMPACTS especially can feel quite alone and frustrated in today's world, though at an earlier time in human development they actually 'ran' the world. Today they provide the fuel on which the world runs.

It is my sincere hope that this book will help IMPACTS everywhere better understand themselves by showing them the derivation of their unique energy and why it is sometimes so difficult for them to find their particular spot in the order of things. I want to give the world another version of reality, one based on what I believe is a more scientific template. This will give people another way to view and assess their own experiences and beliefs. They do not have to settle for the explanations of the current paradigm. I believe there is another, more plausible view.

There is another reason for writing as well. I want this book to help IMPACTS around the world discover new ways to use their precious energy. It has to be utilized; the world needs it desperately.

What makes me qualified to write such a book? Nothing really, but I have learned that credentials mean very little as it relates to knowledge, the quest for knowledge, or the desire to make a difference. It appears that most of the great discoveries and applications of those discoveries in human civilization, including today, were made by people 'on the periphery' with few credentials, people who were somewhat or markedly detached from the structure around them. Einstein was not 'qualified' to write his five papers in 1905 about the universe. He was not affiliated with a university—he was a clerk in a patent office. Isaac Newton was not qualified to discover the laws of motion when he was twenty-three years old, either. He developed almost all of his theories within his own head with very little input from anyone else. Leonardo da Vinci, possessing one of the most brilliant minds ever, disdained organized education.

Thomas Edison had only a few months of organized education before his mother, a former teacher, decided that she could do a better job teaching the curious youngster at home. The school said he was slow and confused; at home he read Shakespeare and Isaac Newton. Plus, he had what you will see often in this book—personal initiative. By age twelve, Edison was selling newspapers, books, vegetables, and candy on the local train, and by age fifteen, he was printing and selling his own newspaper. One day, he acted quickly and saved the life of a boy who was about to be run over by the train. The boy's father was the stationmaster, and to show his gratitude, he taught Edison how to operate the telegraph. Though he had a hearing problem, no college, and very little electrical engineering training, Edison produced one patent for every two weeks of his working life— 1,093 patents in all.

Likewise, James Watt, whose refurbished steam engine made the Industrial Revolution possible, was sickly as a child and consequently schooled mostly by his mother in his early years. He never went to college either except as an employee to work on instruments and machinery. We will see why Edison, Watt, Einstein, Newton, Da Vinci and other great hands-on discoverers, explorers, and inventors are more the norm than the exception. We will also see how the periphery plays an important role throughout the universe as the change agent.

There is a strong thread running through the development of human civilization that asks, "What are the critical needs in the environment at this particular time and what results can be delivered that will optimize the potential?" In other words, what is out of balance and what can be done to bring it into balance? The IMPACTS are that thread and the people mentioned above are some high-profile examples. The vast majority of IMPACTS, however, toil quietly and humbly behind the scenes, content to make the world a better place and seeking no special attention for their efforts.

Why This Book Is Important

As I dug into the characteristics of IMPACTS, I began to see that their profile matched certain principles of the physical world and the universe. As I mentioned, I am one of those people who believes in going back to the first atom, if need be, to find the answers. In this case, I actually had to go back further—to the Big Bang. At one time in this process, I believed that the hydrogen atom was the perfect model for our study, but now I think it is the 2nd Law of Thermodynamics. The hydrogen atom appears to be a representation of the dispersal force (proton) versus the dispersal-hindering force (electron).

I used to think that the more we learned about the universe, the more we would learn about ourselves. Now I believe the opposite—the more we learn about ourselves, the more we will learn about the universe. After all, everything is based on energy, and we are just manifestations of the energy that exists in the universe. But if we are going to learn more about ourselves, we will have to drop time-honored preconceptions and put ourselves back where we belong—in the natural world. Only then will we be able to see with enough clarity to start building a more accurate picture of what is really happening with us, within us, and around us.

We mentioned the hydrogen atom with its proton and electron. Many of us tend to think of the atom as the smallest form of matter but the atom is actually an amalgam of different kinds of matter, principally protons and electrons. Neutrons are an amalgam of protons and electrons. Besides having opposite charges, protons and electrons have opposite energies also. Protons are 'pulling inward' in the nucleus while electrons are in constant motion around the nucleus. This constant motion can be viewed as a captured situation in which the electron may perhaps be trying to escape the clutches of the nucleus. If the outer energy level of the electron field is not full, then the peripheral electron(s)—valence electron(s)—of one atom will hunt for valence electrons of another atom so they can bond together, and in doing so, molecules will be formed.

The manifestation of valence electron energy and other dispersal-hindering energy throughout the universe and of course on earth is what I call IMPACTS energy. It is predominantly female energy—creative, formative, productive, connecting, cooperative and sharing, sustaining, improving, facilitative of change, and duplicative. It is anti-status quo energy. Male energy is usually found around the nucleus whether it is the nucleus of an atom, a large company, the U.S. government, or the black hole center of a galaxy. Its primary concern is in acquiring and maintaining power and control, and this can only be achieved through its capture of innovative-productive IMPACTS energy. I call these predominantly male power-control centers in the human world the SN for structural-nucleus or structural-nuclei.

Throughout the universe and within human affairs, male energy uses this captured IMPACTS energy for its own ends, often with disastrous results for the IMPACTS and the SN in human society. I will discuss the push-pull relationship between these two forces throughout the book. Without an understanding of these two energies, we have no chance of understanding human civilization or human behavior. But here again, it is very simple. Bad paradigms produce unnecessary complexity, and lead to problems that have no solutions. Again, imagine trying to be a scientist in pre-Copernicus Europe. It would have been impossible. The same is true today—the current model makes it extremely difficult to understand the human predicament.

Important points I will emphasize:

1) Simplicity is very important. Let's strive to avoid complications, which often arise when we insert our own prejudices and preconceptions into our research. Renowned American psychologist William James (1842–1910) said, "A great many people think they are thinking when they are merely rearranging their prejudices." In other words, we develop our own reality template and filter out new information that doesn't fit rather than realizing that maybe we need a new or more flexible template.

2) The basic dynamic throughout the universe appears to be structural-nucleus (SN) energy versus creative-formative-productive IMPACTS energy—proton energy versus

electron energy—male energy versus female energy—dispersal energy versus dispersal-hindering energy. Energy organization in the universe appears to follow the template of the atom with a small massive core of 'pulling-inward' energy surrounded by creative-formative-productive 'pushing-outward' energy. The human social model appears to be the same.

3) The creative agent, or energy, generally comes from the periphery, though the periphery can be nuanced.

4) The IMPACTS are the fuel and engine for the human race, and the glue that holds it together. In other words, they and the bonds they share with others power human society, which also prevents catastrophic dispersal or collapse.

5) The future well-being of the world rests with more political (decision-making) power in the hands and minds of the IMPACTS.

As you are working your way through the material, you should begin to be able to identify IMPACTS energy (bonding and dispersal-hindering) and SN energy (dispersal-oriented and pulling-inward) interacting at any moment in time and place. You will see it in your daily life, in the news, in history, in a movie or play, within a book, and everywhere else as well.

How Are We Going to Proceed?

In the first chapter, we begin our exploration of this totally new view of just about everything. Chapter 2 will look at some present-day IMPACTS and their characteristics and how they fit into the concept, including the peripheral aspect. Chapter 3 will delve a little deeper into the theory, and Chapter 4 will discuss the origins of modern humans from the San tribe of Africa. In Chapter 5 the role of the San-shaman and the trance in modern human development will be examined. Chapter 6 will take us out-of-Africa to the development of agriculture, and in Chapter 7 the emergence of the SN or structural-nucleus will be the focus. The contemporary world is the subject of Chapter 8. We will close with a Conclusion.

Throughout the material, the works of other authors will enrich our discussion: Jared Diamond, Stephen Gould, David Lewis-Williams, Stephan Harding, Bill Bryson, Nigel Spivey, Jim Collins, and others. We will discover that the IMPACTS and the IMPACTS concept are very relevant to their writings, and are indeed integral to the understanding of human civilization on every level.

They say the universe is what we say it is. By inference, everything else is what we say it is too. Existing paradigms are defended by those who benefit from their design and construction. The elements in charge try to shape reality to their advantage. Paradigms have 'staying power' because they continue to serve those who control them. Truth doesn't really have anything to do with it. But this is the same story everywhere—the male nucleus attempts to control the structure. It happens within the hydrogen atom with its one proton and one electron, in galaxies with super-massive black hole centers, and on the battlefield where mostly men tear each other to shreds.

Convention says that history is written by the victors. Let's look at another view of the development of human civilization, a view that could have been written by the 'captured'. I think most of us are familiar with the term 'the Stockholm syndrome', named after an event in Sweden in 1973 where bank employees, held hostage by bank robbers for six days, appeared to form an attachment to their captors. How does the Stockholm syndrome relate to human history and the present regarding the control of society by the structural-nucleus or SN? In the Stockholm syndrome, you have an abusive element and you have a captured element that identifies with the abusers. Has human civilization become a day-to-day example of the Stockholm syndrome? As we go forward, you be the judge.

I am trying to do what a scholar tries to do—make you think. Why? Because the world desperately needs alternative views and workable solutions. In the end it comes down to this—what can we do with this newfound knowledge to make the world a better place for human beings and for the earth and all of her creatures? Can we put human civilization on sounder footing? Is it possible for values such as justice and humanitarianism to become ascendant, or have we gone too far in the wrong direction? I do not know, but we will explore the issues.

On this trip, we are going to see the same things we have always seen, but all of it is going to look totally different. It is kind of like going back to a place from your childhood—everything looks the same, and everything looks different at the same time.

I hope this book aids you in better understanding yourself and others and the universe in general. Like IMPACTS do every day of their lives, I am just trying to fill in some of the blanks, and there are many. Our current paradigm does not come close to answering all of our questions and actually leads us astray in many areas. Even its hierarchal structure is a deterrent to our search for answers. We will see that most discoveries emanate from a nurturing circle at some distance from the center of the prevailing structure.

The universe is actually not what 'we' say it is; it is what some of us say it is. The rest of us are still asking questions. Knowledge truly is power because it is a bonding, dispersal-hindering agent with tremendous potential energy. That is why knowledge was kept from the masses so often in recorded history. It is one of the antidotes to the 2nd Law of Thermodynamics among people. It helps keep people together even as dispersal forces try to pull them apart.

Thank you for reading. Your comments are always welcomed and appreciated.

Chapter 1

Getting Started

The goal of science is to find the lowest common denominator that will explain as many phenomena as possible. Einstein's $E = mc^2$ is an excellent example, though he was disappointed that he could not find the one all-inclusive theory that would solve all the riddles of the physical universe.

I think if we are going to solve deeper mysteries of the universe, we are going to have to identify the underlying dynamic(s) in the universe. Presently we are looking at it as if it were a TV show with fireworks exploding now and then (supernovae and meteors), comets passing through occasionally on their long journeys around the sun, and a breathtaking slide show provided by the Hubble telescope. I believe it is much more real and 'alive' than that—very alive as a matter of fact, in that there is a struggle between the dispersal forces of the 2nd Law of Thermodynamics and the anti-dispersal forces. I suspect everything is built around this struggle.

Everything in the universe is energy. That is its currency. The 1st Law of Thermodynamics says that energy can neither be created nor destroyed. It can change forms; e.g., a falling rock changes from potential energy to kinetic energy to heat to sound, but the total amount remains the same. All of the energy unleashed at the Big Bang is all the energy there is in this universe. I say this universe; there may actually be more, with different laws. You never know. That would truly redefine the periphery, but it is not out of the realm of possibility. Anything is possible.

The 2nd Law of Thermodynamics says that localized or concentrated energy will spontaneously disperse if not hindered from doing so. Entropy is the measure of this dispersal. An example of the 2nd Law is an inflated tire. If you puncture the tire with the tiniest pinhole, the pressurized air will immediately start escaping into the surrounding environment. In a very short time, the pressure inside the tire will be the same as that outside the tire. The rubber itself was the barrier that prevented the 2nd Law from being actualized. When those very tight bonds were broken, dispersal ensued instantaneously.

Activation energy is the amount of energy required to break the bonds which are holding back the dispersal forces of the 2nd Law. Not reaching the required level of activation energy allows substances to remain intact and prevents change.

The development of human civilization is following the 2nd Law precisely. There is a powerful dispersal tendency as exists in the rest of the universe but there is a dispersal-hindering element as well. This element is the IMPACTS group I mentioned earlier. The IMPACTS act like the tire—they keep human civilization from dispersing in every direction. But the anti-dispersal bonds, though powerful, are also tenuous because as the structure of human civilization grows, IMPACTS bonding energy gets caught in a tug-of-war between people forces and SN, or structural-nucleus, forces.

In physics, 'potential' is the term used to describe the difference in levels of energy concentration. The 2nd Law wants to destroy potential and disperse all energy until there is total homogeneity. The SN in human society reacts very much the same; it seeks homogeneity but will permit the realization of potential with a condition—that the realization will serve SN objectives or at the least, not threaten the goals of the SN. The SN is a powerful force with a very strong energy field, which is why most people get in line and stay there.

Words can sometimes confuse the issues. An excess of words is often proof that we are groping for answers. Einstein's formula has no words. It contains so much truth that words are not required. Mathematics might be the best way to describe reality, but then who would understand it? Not me. So let's do the best we can with words. And let's keep a healthy dose of common sense on hand, something that often gets cast aside in the scientific world.

For an example of confusion emanating from the use of words, all we have to do is look at the electromagnetic spectrum, which is composed of different energy levels of electromagnetic radiation (EMR). EMR is waves of energy that are produced by accelerating or oscillating or vibrating charged particles such as protons and electrons. Most of EMR is produced by the actions of electrons because their mass is so small they can more easily be accelerated.

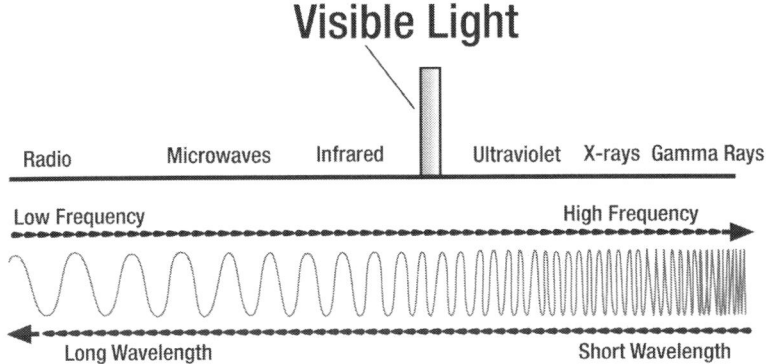

On the least-energy end of the electromagnetic spectrum, we have radio waves. We are all familiar with radio waves on earth but there are also radio waves throughout the cosmos, though they are not sending music. Radio stations generate radio waves and use those waves to carry programming. Moving up the scale, we have microwaves, infrared, visible light, ultraviolet light, X-rays, and then gamma rays. Gamma rays and X-rays are particularly destructive to life's processes. So though all of this energy is of the same kind, it is the differing levels that make it benign or dangerous—to us anyway.

As you can see, in the everyday vernacular, one part of EMR is called light, another part is called waves, such as microwaves, and still other parts of the spectrum are referred to as rays such as gamma rays. If it were a numbering system of some sort, I think we would understand it much better. So let's keep in mind the pitfalls of using words. Often they confuse rather than elucidate. Words can cause us to exaggerate differences and minimize similarities.

Another example of words confusing things is in the naming of the different atomic configurations, or elements. The most common form of hydrogen is 1 proton and 1 electron. Deuterium, an isotope, has 1 proton and 1 neutron in the nucleus and 1 electron. Tritium has 1 proton, 2 neutrons, and 1 electron. Helium-3, a rare isotope of helium, has 2 protons, 1 neutron, and

2 electrons. The most common form of helium, helium-4, has 2 protons, 2 neutrons, and 2 electrons. So the only difference is a little more energy packed into each configuration. Everything is following a natural progression starting with hydrogen.

Names are creating a divide in our minds that causes us to focus on differences instead of similarities. Naming is 'dispersing' the natural connections that we should be seeing. That is why name-calling is used so often in the human world—it is effective at dispersing people. By the way, hydrogen and helium together make up about 98-99% of the regular matter in the universe, with hydrogen being about 73-75% of the total. So you can see that the 1-2% remainder makes quite a difference in the construction of the universe.

The first thing we need to do is go back and start from the beginning. I do not mean we have to relearn everything. Actually all we have to do is look at the same material from a different angle. Much of our confusion and fuzzy thinking comes from the potpourri of what we have been taught, one theory developing from an inadequate other. The complexity we see about us is an unintended ruse, but a ruse nevertheless. It is like the weather; the weather looks terribly complex, and it is on one level, but at its base it too is a manifestation of the 2nd Law—an attempt to disperse the concentrated energy and homogenize the atmosphere.

Some have bemoaned the lack of interest in science among the populace today, especially among the young. But I believe that is partly a result of a lack of true democracy throughout the society combined with an abundance of hierarchal structures. Science is discovery, and discovery requires a lack of boundaries. It also requires the freedom to fail and the freedom to operate without hierarchal dictates. Dispersal forces that promote homogeneity discourage inquiry. Therefore change usually comes from the periphery. It has to because the pressures within stifle creativity and alternative thinking.

Let's not recoil at the challenge of these age-old questions. Often we are more intimidated by expectations of difficulty than we are by the challenge itself. Just because we have been taught or led to believe that certain answers are out of reach does not mean that they actually are. One reason it might seem so is because we live in a hierarchal system that directs inquiry to its needs and away from a natural open curiosity. Our 'free will' is a misnomer;

it actually exists on a continuum. Perhaps we are freer than other species, but an argument could be made against that, too. We have free will within the parameters set for us by forces beyond our control. An alien civilization might compare humans and ants and conclude that both are behaving about the same, and who could argue?

A colony of grass-cutter ants might move 40 tons of dirt while digging 25 feet into the ground in the construction of its nest. The nest will contain fungal gardens for food, garbage pits, and a ventilation system. It takes about 40,000 ants to equal the number of brain cells in one human brain, and there might be several million ants in one colony. The individual ants behave much like cells of an organism, and when joined in activity they act like a super-organism. It is all about connections, just like the human brain. Is it possible that much of the universe behaves the same, as a super-organism? Anything is possible.

The word science instantly makes many people cringe. They relate it to laboratories and physics and chemical equations and all of that. But science is really about knowledge, and learning is a whole lot more fun, and easier too, if one can identify a dynamic weaving itself through the field of study, a dynamic around which everything else is built—kind of like the main character of a novel. As I said, complexity is often a ruse. Let's not get sidetracked by it. Early modern humans focused on the simple in the environment with an eye toward cause and effect, and their accomplishments were prodigious. Of course, they weren't constrained by hierarchal dictates as we are, either.

Sometimes we just cannot see the very simple, the wheel being a prime example. Remarkably, it was not invented until about 5,000 years ago though millions of very smart people had come and gone over the previous 100,000 years and more, lugging possessions over millions of miles. But there is a simple explanation for that also. The predominant energy field of early humans was constructed with powerful bonds, and those bonds resisted change in the environment. It wasn't until agricultural development and the accompanying changes in living patterns started severing those bonds and creating new situations and new needs that new adaptive thinking began to emerge. So just because we have not been able to construct a very simple and valuable model of human culture and its beginnings does not mean that it is not staring us right in the face.

Again, the universe is energy. Energy is not only IN everything—it IS everything. But we still have no clear definition of energy, which shows just how much we really know. Einstein's formula of $E = mc^2$ just tells us that energy and matter are the same thing; it doesn't tell us what energy is.

We may not be able to define energy adequately, but we know that there are different forms of it. The energy of a hurricane is different from the energy of the sun on a hot summer day. The energy of a mother toward a newborn child is different from the energy of a soldier going into battle. The same principle applies to the hydrogen atom or any atom; proton energy is very different from electron energy. The atom is a marriage between these two energies; most structures in the universe appear to follow this same basic model, even the institution of marriage between people.

The atom appears to be the 2nd Law and the 2nd Law-breaker 'personified'—the dispersal force of the proton is counteracted by its antidote, the electron. A nucleated structure such as the atom is a hybrid or compromise between the two major forces, each diametrically opposed to the other. The proton is male energy— the electron is female energy. When the male proton 'marries' the female electron, the female electron often bonds with electron energy from another atom, producing a molecule. Molecules hold energy within the chemical bonds and prevent it from succumbing to the 2nd Law. When human males and females connect and have children, the same thing occurs—the creation of more humans continues the bonding needed to sustain this life form and its energy, preventing dispersal.

In the human race the lines between the two forms of energy get blurred somewhat but still, at the foundation, females can create new life and males cannot, just as electrons can bond and form molecules and protons cannot. In many life forms, males are used almost exclusively as sperm donors in order to diversify the genome and make it more adaptable to the environment. But males have carved out a more prominent role for themselves in the human species.

It is no accident that men usually start and participate in wars while women are more likely to try to prevent them, even as they alone can provide, through birth, the human ammunition that enables them. The epidemic of tragic killings by youthful males in America is not a total surprise either. Young girls do not usually go around killing other people. Destruction is part of dispersal, and dispersal is part of male energy. What prevents dispersal? Strong bonds, something that regretfully many young males do not have today. Strong bonds 'civilize' dispersal tendencies.

It is the story of the universe and the nature of the two energies. One is creative-formative-productive energy—the other takes that energy and uses it for its own ends, often to enlarge its domain through capture and/or destruction. The phenomenon is universal; e.g., a galaxy will often gobble up another galaxy. Businesses do the same, countries, too. Like the atom, they combine both energies with the structural-nucleus or SN in control. What gets our attention around the globe is extreme SN male energy. It grabs the headlines and everything else that it can. It is a powerful pulling-inward force, like a black hole, that leaves destruction in its wake. It is also extremely difficult to keep under control. You will note that when one extreme-male-country overreaches and loses its place on the world stage, another alpha quickly replaces it.

We are going to continually see that violence usually emanates from the 'male nucleus' or near the nucleus, whether it be cosmic rays which are predominantly protons, gamma rays which mostly result from electromagnetic disturbances within the nucleus, or X-rays, much of which originate from electrons close to the nucleus. Gamma rays and X-rays are generally produced around violent events such as a supernova. A supernova sometimes occurs at the end of a star's life when the fuel has been exhausted, thereby gravitationally causing it to collapse on itself, resulting in a terrific explosion. Also, large amounts of antimatter in the form of positrons have been found in the vicinity of black holes, and in proximity to binary stars where a dying star is accreting gas from the other healthier star. A positron is the anti-particle of an electron. When electrons and positrons collide, they are destroyed and gamma rays are produced.

The accreting entity appears to hold the most potential for violence, probably a result of its inability to create its own energy, or enough energy to satisfy it. If an entity does not create or produce its own energy, it must accrete it. Creation and bonding are on one end of the spectrum; destruction and dispersal are on the other. All of these universal physical events have applications in the human world as well. Everything is operating according to the same template, as we will continue to see. For example, females are creating and developing bonds with new life, and males, all too often, are destroying each other, both domestically (within the country) and on the battlefield.

Electrons make just about everything on earth work— machinery, technology, photosynthesis, and molecular formation, including DNA and water—just to name a few. Of course, water and many other molecules are found throughout the universe. Without molecules in molecular clouds of hydrogen gas, there would be no stars, and no light of any consequence, and certainly not the life we know. Electrons are the workhorses. They appear to be the transformative agent, certainly on earth, but also throughout the universe.

Einstein's formula obviated the need for incessantly difficult calculations and presented the basics of the universe in a new light. Energy and mass were two sides of the same coin; the only difference was the form. Can we find a formula that describes the development of human civilization? I think we can and I think you will find it to be very simple, just as Einstein's formula is simple.

In an atom, the number of positive protons equals the number of negative electrons. But still things are not in balance except in a very few atoms called the noble gases—helium, neon, argon, krypton, xenon, and radon. In the noble gases, the outer electron energy levels are filled with the 'right' number of electrons. In any atom, the energy level closest to the nucleus wants 2 electrons, the next wants 8, the next 8, the following 18, then 32, and so on. The atom, or maybe it is the electron field itself, is not going to be happy until the outer energy level is full. If it is not full, the electrons in the outer energy level are called valence electrons. These valence electrons are ready and willing to bond with valence electrons of other atoms, forming molecules and compounds. Most of the time, electrons do this by bonding in pairs. Pairing is ubiquitous in the universe. For example, most stars are binary stars, not single as is our sun.

Oxygen has 6 valence electrons. If two of its 6 electrons can each bond covalently with the electron of a hydrogen atom, forming a molecule of water or H_2O, then all three atoms will be satisfied with the electron configuration in the outer energy level. One hydrogen atom will have 2 electrons, the other hydrogen atom the same, and oxygen will then have 8 electrons in its outer level. So a valence electron and its desire for bonding with another electron are at the base of what drives the production of a more complex atomic structure. The change agent resides on the periphery. It forms the bonds that work against dispersal.

Oxygen – 6 Valence Electrons   Hydrogen – 1 Valence Electron

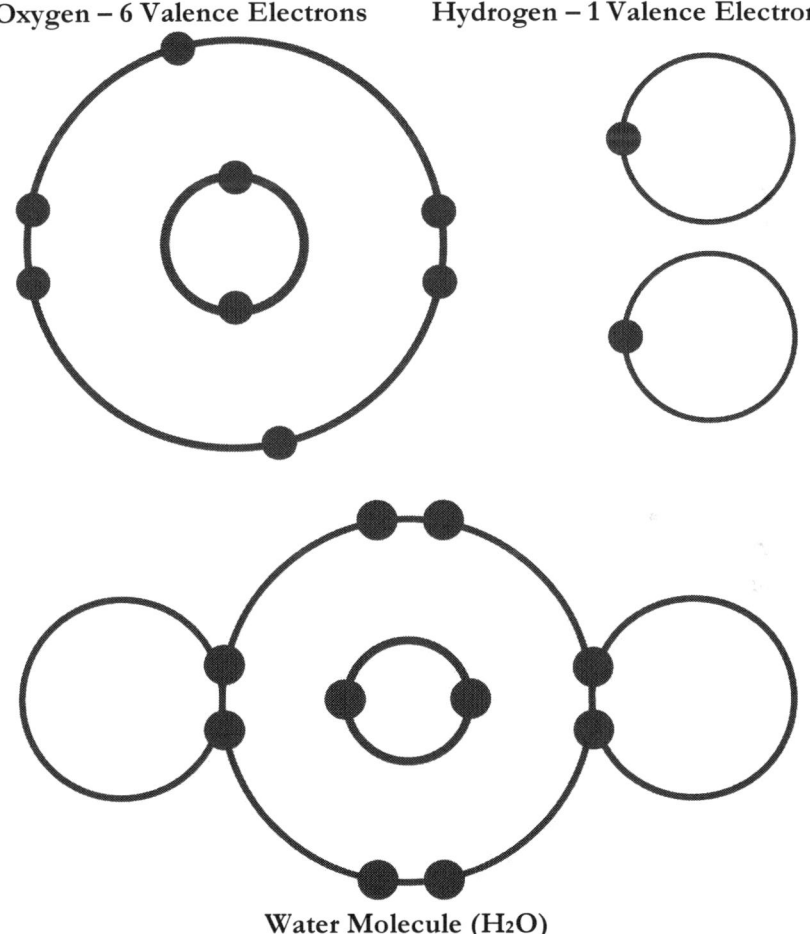

Water Molecule (H₂O)

Human beings as a group do not act exactly like the atom but the model is precisely the same. We are just bigger and slower than atoms, though the forces are identical. Throughout the universe it appears to be the same two forces—one pulling fiercely inward, the other creating, forming, and producing, mostly from the periphery. There has to be a balance between these two pressures or it appears that collapse is the result, as happens with a dying star that has used up all of its fuel. The lack of fuel means the cessation of outward pressure.

In human society and civilization, we are going to find the exact same two forces. As a matter of fact, that is the story of human civilization since the development of agriculture—the struggle between the pulling-inward force that I call the SN, or structural-nucleus, and the creative-formative-productive force of the 'human valence electrons', the IMPACTS. The IMPACTS are the fuel that has to continually burn or else the society collapses. Dark ages occur when the IMPACTS' energies have been destroyed, muzzled, or dissipated.

Remember the illustration of the tire and the puncture hole with the air rushing outward and diffusing? The chemical bonds of the protective structure were broken just enough to render the structure ineffective. It didn't take much. The same happens in dark ages as human bonding gets severed and IMPACTS energy loses its cohesion. And as we know from history, it can take considerable time to recover from such events. The current financial meltdown (2008-2009) is proof of how tenuous the modern financial structure really is. Economic, societal, and personal bonds have been shattered along with bonds of trust and confidence. IMPACTS have withdrawn to reassess the situation, and it will take some time for them to reengage. As they do, the economy will gradually improve.

Atomic Capture

I have used the word capture to describe the dynamic operating between male energy and female energy. But I have seen the same terminology used again and again in my research when others are discussing the formation of the first atoms—the word capture is used almost universally to describe the marriage of the proton and electron with the proton doing the capturing. It was probably the first instance of female capture by male energy after the Big Bang. It doesn't appear to be a marriage made in heaven, not to me anyway, after all my research and writing.

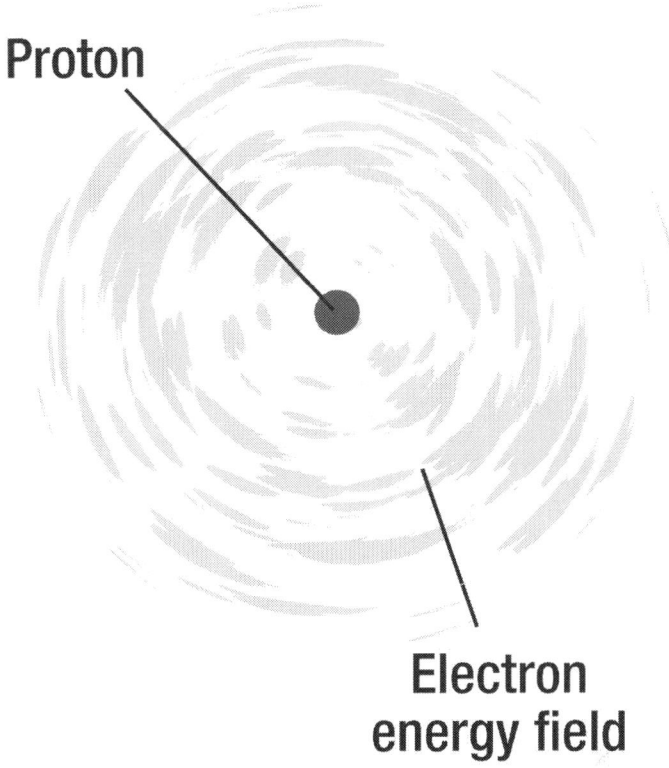

Proton

**Electron
energy field**

Is the Electron Captured?

We have been taught that the proton and electron are attracted to each other because of opposite charges, but Einstein's general theory of relativity may have something to do with it also. Part of the theory states that mass or energy bends and distorts space-time around it, depending of course on the 'massiveness' of the mass or energy. So it would make sense that the massive proton or nucleus could influence its surrounding space-time, in effect capturing electrons beyond the positive-negative attraction.

Let me give you an example of how others view the early formation of atoms. George Musser states in a recent <u>Scientific American Magazine</u> article:

"First, it took a while for protons to get a firm hold on electrons. Their grip was tentative at first. To tighten up, a newly formed atom had to lose energy by emitting photons, and it did so in its own good time. Further complicating matters, a photon from one atom tended to knock the electron off another. Like crabs in a bucket, atoms thwarted one another. What overcame their mutual antagonism was cosmic expansion, which sapped the photons' energy and gradually tilted the balance in favor of atom formation over destruction. The commonly cited time frame of 400,000 years is just a convenient milestone; recombination (atom formation) actually took as long as a couple of million years to run to completion."

You can see that the protons only got a strong grip on the electrons through cosmic expansion, which is a form of dispersal. This further broke the cohesion or potential cohesion of electrons, and strongly mitigated their ability to shape events, placing them in service to the protons. The same happens in human society—the SN tries to make sure that IMPACTS energy is not concentrated in the 'wrong' areas and is concentrated in areas that will prove useful to the SN. Political protest is the 'wrong' concentration—a technology start-up is the right concentration.

Mr. Musser appears to showcase the long struggle of atom formation as an issue mostly between atoms—"atoms thwarted one another". But I think it was a contest between protons and electrons. Any way you look at it, it doesn't look like a benign union, does it? It appears that the electrons were pulled to the altar. As we will see with my study, the early formation of human civilization followed the same model with the IMPACTS being pulled (accreted) into the SN 'atom'—or destroyed if they chose not to come willingly.

Herdsman Ants and Captured Mealybugs

In the Malaysian rainforests, a symbiotic relationship has been going on for millions of years between herdsman ants and mealybugs. The mealybugs chew on the leaves of certain plants and willfully share their honeydew with the ants. The ants in turn protect and care for the mealybugs, picking them up and carrying them to other leaves and trees when new sources of food are needed. The herdsman ants also accept the mealybugs into their nest, aiding them in the reproduction and care of their young. The mealybugs have been assimilated into the ant colony. Two different species have combined in a process of survival though it is obvious that the mealybugs are providing the fuel for the relationship.

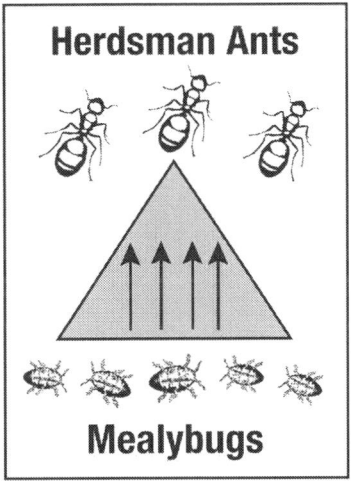

Mealybugs Are Supplying the Ants with Energy

Yes, the relationship is symbiotic, and each side is very content with its specific role. But this relationship developed over a very long time. We can be sure that there was a struggle for quite some time, many mealybugs not wanting to be subservient and share their food with ants. But natural selection prevailed and produced today's relationship. Those mealybugs who did not share their food gradually died out as the ants became the ruling SN of their domain. This domain grew bigger and stronger over time with the help of the mealybugs who were subservient. But accretion is a normal part of nature and the universe. The herdsman ants and mealybugs are just another version. Sometimes it is benign and friendly, sometimes not.

As we will see, the same dynamic has been at work clearly and visibly in human civilization over the past 6,000 years or so. The IMPACTS have become the captured mealybugs to a very large degree though the process is still underway. And yes, it has been a struggle too and continues to be because all of the IMPACTS have not gone quietly. If we ever reach that state, there is little hope for the current human species. Why? Because the IMPACTS enable humanity to adapt to new conditions of any sort, and for this role they must possess an anti-status quo nature. The SN of humanity tries to control the expression of this nature, using it in measured ways.

The herdsman ants appear to understand their situation far better than we do ours, even though an ant's brain is 1/40,000 the size of ours. How does one account for that? Obviously, brain size has little to do with awareness of situational dynamics. Our main problem is that a small percentage of the SN population makes human life precarious for all. Human beings have not yet acknowledged this, much less found a way to deal with it in a practical manner.

The ants know that the mealybugs are their sole source of energy, and they respond accordingly. The SN still does not know the origin of its energy, though it generally believes it is propagating it with its 'enlightened policies'. That would be like the ants asserting that they were responsible for the energy that was created because they were the ones carrying the mealybugs from tree to tree. The transportation part is true, but the mealybugs got along quite well long before the ants arrived. The exact same argument applies to the SN and IMPACTS relationship—we will learn that the IMPACTS were here for 100,000 years before the SN emerged and took control of the energy supply.

Quantum mechanics, or quantum physics if you prefer, is the physics of the atom. Much of the theory concerns the behavior of electrons, including valence electrons, and how they behave under confinement or capture. On the macroscopic level, we don't see the physics of quantum mechanics; it is only evident at the atomic and subatomic level. Civilization is the same. We only see it macroscopically; we see the clash between and among cultures and countries. In actuality, the most important part of the picture is occurring underneath in the dynamics between the SN and the captured IMPACTS within a particular country or empire or group. You could call the SN-versus-IMPACTS struggle the quantum mechanics of human civilization.

If you had a war between two colonies of herdsman ants, what would be the most important variable? It would be the relationship between the ants and the captured mealybugs on each side. It is the same in the human world.

I suspect that the musings of post-agricultural philosophers are very different from those of pre-agricultural philosophers. For one thing, philosophy before agriculture included everything under the sun—science, medicine, metaphysics, astronomy, the entire universe. It was not really philosophy as we know it; it was scientific inquiry. It was focused on the WHY of everything, and the applications of the answers. But early humans were not captured by an SN; the SN did not develop until after agriculture. They were not confined. There was no ceiling limiting discovery. The 'sky' was the limit. As SN dominance spread, boundaries and barriers appeared in almost every human endeavor. So now philosophers talk only about philosophy, or do they? I suspect, since most philosophers are IMPACTS, what they are really pondering is their state of capture.

The SN-dominated world has essentially chopped everything into little pieces. Again, this is part of the 2nd Law of Thermodynamics, a dispersal of concentrated energy. That is what a hierarchal structure does—it reorients the existing dynamics left over from the creative-formative-productive beginning.

What we will continue to see is an accretion aspect throughout the universe just as we see in the atom and in the herdsman ants-mealybugs relationship. It is male, and it is part of dispersal. If concentrated energy is being taken and broken down for use, anti-dispersal bonds are being broken. This accretion element occurs ubiquitously in human civilization. One force is creating, forming, and producing, doing the hands-on work, and the other is taking control of the results. You see it in personal relationships, in business, in politics at every level, between and among countries—everywhere.

Slavery and colonialism were forms of accretion, but today people and countries find more 'civilized' ways of extracting the energy they want and need. But that does not mean that exploitation is less pervasive. Actually it is probably more so. Now it is hidden behind the sanctity of the state and relations among states. The various SNs around the world enforce the rules in addition to making them. Murder and theft can now be described as self-defense by a country. People fighting for justice can be labeled terrorists when the real terrorists may actually be the leaders of a country or countries. Statehood provides a protective barrier that allows many of the truly 'bad guys' to hide from responsibility for their actions, and to literally get away with murder.

Part of the story of civilization since agriculture can be viewed as an attempt by the IMPACTS bonding forces to 'civilize' and rein in the SN, which is not what the IMPACTS were bred to do. As we will see in Chapter 4 with the San tribe of Africa, the IMPACTS originated to solve problems, beginning with individuals who were sick. They would have had no way to know that their skills would be needed to fix sick societies. Today's world of unending warfare demonstrates how difficult the task continues to be for the IMPACTS, and how little progress has been made. The SN wants us to accept that this level of destruction is normal when in reality it is way out of balance.

Protons and Electrons

We generally refer to protons and electrons as particles of energy, but they are really a wave-particle duality. That sounds confusing to us only because of the present orientation of our consciousness. At an earlier time, it may have made perfect sense. String theory has recently come on the scene. It suggests that matter is made up of vibrating strings that exist in perhaps eleven different dimensions compared to the three dimensions with which we are currently familiar. Most important to our discussion are the two different energies, the male-dispersal proton energy versus the female-dispersal-hindering electron energy.

Here are some important elements of the atomic model to keep in mind as we explore further:

- Proton energy and electron energy are very different in nature and behavior.
  - Proton energy is massive pull-inward energy.
  - Electron energy is light, mobile, connecting, and creative-formative-productive.
- Electrons seem to prefer to be paired. If they are not, then they are out of balance and hence, the whole structure

(atom) is out of balance, even though the charges may be equal.

- The peripheral valence electrons have more energy than do the close-in electrons because they are farther from the nucleus. Hence they can 'slip away' and connect with valence electrons of other atoms, thereby changing the structure of the atom by forming molecules.

We can tell a lot about energy by where it is located in the structure. This statement will hold true not only in the atom but in human structures as well. Because it is creative-formative-productive energy (CFPE), we will find IMPACTS energy around 'new beginnings' such as the formation of a molecule, births, entrepreneurialism, immigration, invention, exploration, discovery, artistry, writing, social movements, new ideas, and more. IMPACTS energy 'brings into being'.

As we will see, creative-formative-production usually takes place in a close circular environment such as exists with the electron, or with a family. But when the development reaches a certain point, a generally male-nucleated-hierarchal entity emerges and takes control of the structure and the female CFPE, or IMPACTS energy. Or we could say the nucleus of the structure accretes it. We see it in businesses and organizations that usually start with a small circle of boundless and boundary-less energy, which ultimately settles into a CEO and board-directed management. We see it in political revolutions too as the revolution gives way to a hierarchal government. Unfortunately, sometimes we see it in families in which the male actually captures the female, literally and/or figuratively. This practice still occurs frequently in many parts of the world and has been prevalent throughout recorded history. The capture, or accretion if you prefer, of creative-formative-productive energy is a recurring theme in the natural and human worlds.

The IMPACTS

The valence electron energy level continues to be an excellent model for understanding IMPACTS energy and therefore IMPACTS people. It appears to be the primary bonding agent in

45

the material world and hence the primary anti-dispersal agent. The valence electron is no different physically from the inner electron; it just has more energy because it is farther away from the nucleus. It is also in an energy level that is not filled with the preferred number of electrons, and is therefore 'motivated' to correct the imbalance.

How the Valence Electron Compares to IMPACTS:

- The intent is to connect with other electron energy which will reestablish balance. Using words to try to explain the actions of the valence electron, we could say that the valence electron is searching for opportunities that will improve the situation and the existing structure. We will see that IMPACTS have always had the same philosophy.

- The actions that would achieve this reestablishment of balance would be the actual connecting and bonding with other electron energy. IMPACTS have always practiced this also, trying to connect and share ideas with other IMPACTS.

- The results of valence electron bonding would be a molecule which has been created and formed, and which has consequently changed and enlarged the structure. In the process, the creative-formative-productive energy has become embedded within the new structure and is now integral to the structure, but not necessarily in control of it. It is the same with the IMPACTS and human civilization; the IMPACTS and their innovative energy are embedded but they are not in direct control of the structure. Just as electron bonds store energy, so too do IMPACTS' bonds with other IMPACTS and other people store energy, which can potentially be released in a myriad of ways.

- The initial part of the process can be seen as 'mining' by the valence electron as it is searching for a solution. IMPACTS too are always mining the environment for anything that will make a difference.

If we understand what is happening in this little corner of the atom and how its actions are creating bonds and hindering the 2nd Law, we are well on our way to understanding IMPACTS energy and its manifestation within IMPACTS people. It can be very simple if we will unclutter our minds of previous teachings and assumptions.

Why were the IMPACTS captured? To solve problems and to provide a multitude of products and services built around innovative, artistic energy. This IMPACTS energy has tried to keep human civilization moving forward for the past 6,000 years even as the SN has moved it backwards in many ways. CFPE has little in common with SN-structural energy just as the electron has little in common with the proton. The SN and the IMPACTS are working at cross-purposes, and what we call civilization is the result. Protons and electrons are doing the same with the atom as the result.

IMPACTS energy can best be thought of as analogous to a mother's love of her newborn child. No matter where you look in the animal kingdom, you see the same general maternal behavior—creative, formative, productive, hands-on, innovative, nurturing, and protective. It is hard to match the ferocity of a mother protecting her offspring. That is passionate IMPACTS energy—protection of bonds deemed to be extremely valuable. You see this passionate energy manifested by people in many ways: toward those in need, including animals and the earth itself, in business development, through involvement in communities, in families and friendships, through the practice of religion, and in countless other ways. At its base, it is all about strong bonding—anti-dispersal bonding.

The SN, which is predominantly male energy, wants to capture IMPACTS for the same reasons that a man wants to 'capture' a wife—there are many potential benefits including energy that can duplicate and therefore expand the structure. This can translate into power, and possibly greater control. IMPACTS energy, as evidenced by the mother's love for her children, emanates from deep caring. But this energy in the wrong hands can become warped and deadly. Today's world is the evidence. The SN has captured IMPACTS energy to be used as it pleases. There are wonders galore and there are horrors galore.

The contemporary world is an SN and IMPACTS hybrid, just as the atom is a proton-electron hybrid. Those who possess both forms of energy can be quite 'successful' in navigating the modern world. Those with an abundance of IMPACTS energy and little SN energy can find rough sledding. It is not their world; that is why we find so many IMPACTS on the periphery.

IMPACTS energy is like athletic ability or art ability—we all possess varying amounts of it. There are not many Michael Jordans or Picassos out there but there are many gifted athletes and artists. So we are going to see varying levels of IMPACTS energy in the general population.

I treat human civilization as going back about 130,000 years or so when homo sapiens sapiens emerged. (Yes—two sapiens.) Homo sapiens came on the scene about 200,000 years ago but gradually evolved into homo sapiens sapiens, today's version—us. The behavior of homo sapiens sapiens appears to have changed little until the advent of agriculture about 10,000 years ago. Agriculture changed everything because with a food supply assured, humans could form permanent settlements and societies, which meant the conditions were right for hierarchal structures and SN leadership. This in turn created the necessity for new roles that served the structure and the SN: bureaucrats, full-time craftspeople, scribes, servants, farmers, laborers, weapon designers, warriors, and more.

To service a structure, you need ever-present energy, and different kinds of it. Mostly you need loyal, hardworking people combined with innovative energy. IMPACTS energy manifested in IMPACTS individuals became the primary source of innovative energy for the emerging societies and civilizations. With innovation needed across the board, the IMPACTS became the glue holding everything together—economically, culturally, socially, and scientifically.

In the manner in which I have defined them, I believe the IMPACTS are a small percentage of the human population, probably around five to ten percent. But they have a large support group, without which much of their innovation could not be utilized by humanity. In some areas the population density of IMPACTS will be greater than in others, but IMPACTS will be found wherever human beings are found. The IMPACTS are not an elitist group and I do not mean to infer that they are. Most are very regular people and would strenuously object to being characterized as special in any way. They are no better or worse than others; they just get things done. Some IMPACTS are nice people while others are not. Some have a highly developed conscience; others have little.

Someone asked if IMPACTS weren't just those near the top of the class. Some are; some aren't. Being an IMPACTS-person has nothing to do with rank in class, SAT scores, or college graduation. Rather it is a drive to make things better which can be housed in any person anywhere in the world, from the poorest peasant to the richest entrepreneur. But let's not forget that the goal of the SN is to bring as much IMPACTS energy as possible within its domain—to embed it in an SN-IMPACTS hybrid structure. And that goal has been largely realized around the world. Therefore, IMPACTS energy has lost much of its purity and is now in service to the SN.

Sometimes IMPACTS create, form, and produce a product, service, or structure based upon a need in the environment. Other times, they position themselves within a structure at a critical point, which will often be the most vulnerable position where a meticulous, steady hand is required, such as in accounting. Frequently, they will be project managers or team leaders. But since IMPACTS energy today is usually captured energy, innovation in service to the SN can return healthy profits but sick results. Did you know that the same man, Thomas Midgley, Jr., invented leaded gasoline and chlorofluorocarbons (CFCs)? This is not unusual in today's SN-IMPACTS hybrid world. How do you think such horrific weaponry gets invented and produced? The SN captures the needed, innovative IMPACTS energy.

Clearly we are talking about energy; people are carriers of the energy just as the atom is a carrier of electron energy. Some people with less IMPACTS energy actually contribute more to the world than those with copious amounts because of what I mentioned earlier—the SN is going to take what it can and will sometimes use it in a diabolical manner. If the SN is extremely formidable, as is the case with the U.S. societal structure in general, one thing is certain—large amounts of IMPACTS energy are in service to the SN.

The oil companies, automakers, and air conditioner people loved Mr. Midgley and made a ton of money from his work, which translated into power for them and the government, which also made a great deal of tax money on all of it. Sometimes IMPACTS energy in service to the SN turns out well for society—sometimes not. Conscience far more often resides on the IMPACTS side of the equation than on the SN side. The SN wants 'make it happen' IMPACTS energy primarily—it does not want the conscience part that might come with it. Now you are seeing why we have the world we do; the SN is firmly in control.

Neither leaded gasoline nor lead in plumbing was outlawed in the U.S. until the 1980s. It was the early 1990s before the use of lead was discontinued in canned food products. But the Roman engineer and architect Vitruvius had already warned of the dangers of lead in contact with drinking water in the first century BCE. Such is the ongoing battle of the IMPACTS and common sense against the intransigence, and worse, of the SN. The SN has remained basically unchanged since its emergence after the development of agriculture. But that makes sense—as mentioned, it is the valence electron which changes the atom, not the nucleus. It is the same in the human world with the SN and the IMPACTS.

If the SN can get the IMPACTS within its sphere, then perhaps it can channel their dissatisfaction with the status quo into innovation that will strengthen SN power and control. But it will not get them all. Many will stay on the periphery and express their anti-status quo attitudes in other ways.

When you see impressive physical works such as a newly designed hybrid automobile, beautiful and interesting artwork, cutting-edge technological innovation, or a sparkling new transportation system, you can be certain that IMPACTS and IMPACTS energy were involved in significant numbers. The same holds true if you encounter abundant social good such as a teacher helping special needs children, a fundraising walk for multiple sclerosis or breast cancer, the recovery of a blighted downtown area, or the humanitarianism accompanying a natural or man-made disaster. In many ways, IMPACTS are like white blood cells that rush to the injured area and work to stave off infection. Their actions, whether they know it or not, are intended to create bonds that will hinder dispersal, and to preserve those bonds that already exist. The IMPACTS try to ensure that no one is left behind, including animals that need rescuing. It is all part of making everything and everyone well and healthy—being inclusive and creating the proper balance. But when you see a new smart bomb or a weapon that disintegrates a human being on contact, that too is IMPACTS energy—captured in service to the SN.

A dearth of good often accompanied by extensive pain, such as the presence of war, stalemated political haggling over healthcare, or poverty in the midst of affluence, is evidence that IMPACTS energy is in short supply or is being used inappropriately, and that IMPACTS individuals of conscience possess little power or control over the situation. IMPACTS solve problems, so if the problems are of a long-term or seemingly intractable nature, you can be confident that walls have been constructed in order to keep the problem-solving abilities of the IMPACTS out. This is when you can be certain that the SN is firmly in control.

In the next chapter we will look at some IMPACTS in today's world so we can start to get a picture of how human societies actually function—how the IMPACTS provide the innovative-productive energy and the SN captures, manages, and uses it, not always with positive results.

Summary

The 2nd Law of Thermodynamics and its tendency to continually disperse energy appears to be the major force in the universe, and everything that we see of a structural nature is defying the 2nd Law. The atom appears to be a small-scale model of this battle, with the protons as part of the dispersal force and the valence electrons hindering dispersal. They do so by forming bonds with the valence electrons of other atoms.

Valence electrons form bonds throughout the universe, especially on earth. Without valence electron sharing and bonding, life as we know it would have no chance to take hold and develop because that is what life is—a conglomeration of molecules. Nor would there be any stars because stars are formed in molecular clouds of hydrogen gas. Everywhere you look you see electrons and their actions holding the universe together, or maybe I should say, preventing it from falling apart.

In the people world, the IMPACTS are the valence electrons that form the strong bonds that attempt to hold humanity together against the SN forces of dispersal. The bonds of the

IMPACTS in society must be flexible enough to permit positive change and strong enough to prevent negative change. Flexibility is not a problem; preventing negative change is extremely difficult.

As the valence electrons are the change agents for the atom, so too are the IMPACTS the change agents of human society. They seek to balance and improve human civilization, and they do it through their own creative-formative-productive efforts and energy. But the SN stands ready to capture, or accrete, all usable IMPACTS energy within its domain—and to use it to enhance itself.

Chapter 2

Real World IMPACTS

Most of the following discussion will be about visible movers and shakers, but do not be misled by that. As stated previously, the vast majority of IMPACTS live quiet lives, performing their work mostly behind the scenes, not seeking or receiving special attention, which suits the SN perfectly. The SN wants nice, quiet 'mealybugs' who just do their work. Plus, there are really two general categories of IMPACTS. One group is closer to and more involved with the mainstream of society; let's call them the SN-IMPACTS. These IMPACTS may come from a nuanced periphery inside the main structure of society, as we will see with Bill Gates and Steven Jobs. Let's keep in mind that every structure, no matter where it is located, has a periphery.

The other IMPACTS group is closer to the periphery of society—the P-IMPACTS. This arrangement follows the atom where, as it grows in atomic number (number of protons or electrons), much of the electron energy is close to the nucleus, as are the SN-IMPACTS, and some is farther out on the periphery. These valence electrons have more energy and are less affected by the nucleus, very similar to the P-IMPACTS in behavior. P-IMPACTS often live their lives paying little attention to the SN, or maintain their distance because of their contempt for it.

Warren Buffett, the investment icon, is a good example of the SN-IMPACTS. He clearly has abundant IMPACTS energy and resides comfortably within the mainstream. With a net worth of $40 billion or so, he pays himself only about $100,000 annually as CEO and chairman of Berkshire Hathaway, his investment firm. Buffett continues to live in Omaha in the same gray stucco house he purchased four decades ago for $31,500. He eats burgers or steaks for lunch and dinner, always washing down his meals with Coca-Cola, produced by the company in which he has invested since 1988. If you had invested $10,000 in Buffett's firm in 1965, you would be worth over $50 million today.

Buffett is a 'regular guy' with uncommon achievements, and now he plans to give the results of those achievements away in an alliance with the foundation started by Microsoft founder Bill Gates and his wife Melinda, other examples of SN-IMPACTS. The Gates have recently focused their efforts on a cure for malaria and on educational improvements in the U.S. Health and education are strong IMPACTS concerns. It is pretty obvious that Warren Buffett was not achieving for himself; he was focused on the greater good and on his clients. Through his acumen and hard work he has benefited many: individual investors, companies which received his investment funds, the local economies and communities where those companies are located, his own local economy and community, the national economy, and the international economy. And now he will aid many others in areas near and far. Not only has he demonstrated that skill can pay big dividends, but he has also shown that service to others can be an important part of the equation.

Buffett, displaying the strong protective element and acute insight that we see in many IMPACTS, was one of the few who were concerned years earlier that the variables were in place for the financial crisis that occurred in late 2008 and early 2009. He believed that risk was being accumulated in far too few hands while many 'experts' preached that the market was its own best policeman. This is something we need to remember—the SN often does not pay attention to the advice of the IMPACTS until something breaks—badly. Then they will have to rely on the IMPACTS to fix it, if it is fixable. This occurs mostly behind the scenes as IMPACTS come up with innovative ways to deal with the new environment.

I mentioned Bill Gates. Before Bill Gates dropped out of Harvard and made himself known around the world as the founder of Microsoft, his parents were nurturing not only their children but also their community, as IMPACTS are wont to do. His mother was a schoolteacher and very active in the United Way, and his father was an attorney who volunteered extensively for Planned Parenthood, hospitals, libraries, and the United Way also.

Gates was a programmer by choice and by nature. Programming is not for everyone; it requires patience and precision as gibberish is transformed into the usable and practical. IMPACTS often do just that. They take something that appears to have little value as is and then develop the hidden potential—in people, organizations, businesses, communities, churches—you name it. They are miners, constantly looking for the vein of under-utilized value, leaving nothing behind. As noted in the Introduction, it was a steam engine that James Watt refurbished that enabled the Industrial Revolution, not a completely new invention.

IMPACTS are masters of frugality; they have an uncanny ability to find value in just about anything. They often transform the environment through invention while at other times they are rebirth and renaissance specialists. It is all about taking what the environment gives and getting the most out of it. Why? So production of positive results can continue. Positive production aids people in solving problems, which helps keep interpersonal bonds from unraveling and causing dispersal.

The reason we see IMPACTS involved with the latest technology and anything else that is cutting-edge is because of the creative-formative-productive element, the 'new beginnings' factor. And that stems from their recognition of needs in the environment, imbalances that they think they can help correct.

SN-IMPACTS can originate on the far periphery and make their way into the center, or they may come from a peripheral area inside the mainstream. With Bill Gates, his decision to drop out of Harvard and pursue the development of Microsoft was not a society-sanctioned approach—until he became successful. IMPACTS are very often taking the opposite track from what the SN prescribes because the SN is about the status quo. IMPACTS are about change which means getting off the well-traveled path in the search for answers to critical needs.

Gates appears to take to the limelight more so than does Buffett, and to have more expensive tastes, proof that SN-IMPACTS come in all shapes and sizes. Gates spent $100 million or so on his house. Most IMPACTS are humble and under radar—some not so much. Some exhibit a power-control element—some have none of it. Some are poor, others are rich, but most are in-between. The vast majority of IMPACTS, no matter where they are located on the IMPACTS energy continuum, pursue excellence as they define it, and generally possess a determination to make a difference.

Regular Everyday IMPACTS People

As noted, most IMPACTS are regular, everyday, hardworking people. One such person is Paul who works at a major home improvement retail company. Paul's ethnicity is multicultural— French, Spanish, Dutch, and Mexican. This is seen often with the IMPACTS; it shows their borderless philosophy and their ancestors' openness to other cultures around the world. IMPACTS often view themselves as citizens of the world more than citizens of a specific country.

Paul grew up in El Paso, Texas, but was always fascinated with the sea. First he was a life guard, but when he turned 18, he joined the Navy and became a corpsman. A corpsman is a hands-on nurse, serving the Navy and the Marine Corps. Attending college after service, he majored in Business Administration and Industrial Supervision. Paul worked at a shipyard for a few years, then in systems analysis for a major computer company, in oil and gas, and then for a major airline where he held several positions. Upon retirement, Paul joined the home improvement company—the next day—and has handled many different responsibilities there as well. IMPACTS never really do retire. Paul is always fixing up the house and doing construction-type favors for friends and acquaintances, usually for very little cost.

Interestingly but not surprisingly, two of Paul's three wives have been emergency-room nurses. Why would that be the case? IMPACTS frequently find partners, marriage or otherwise, who see the world pretty much as they do. The IMPACTS attitude is to be alert for critical needs in the environment and ready to provide solutions.

There are other IMPACTS employed at the same store where Paul is working. Susan, an avid reader and African-American, works in the contractors' section and is a single parent of two teenagers. Her son, AJ, is also an avid reader. The oldest, Linda, just received a scholarship from a Christian college out-of-state. Her chosen field of study? Critical-care nursing, just as we saw with Paul and his wives. IMPACTS often position themselves at a precarious point where things can go very wrong or they can go extremely well. Most people run from responsibility; the IMPACTS run toward it.

Ann is recovering from a bout with leukemia which took her out of action for over a year. Always upbeat and positive, she exudes love of people and life. Her youngest son Adam, age 25, recently published his first book, which was picked up by a major publisher, and her oldest son just returned from working in a war zone as a civilian.

Sean is in charge of inventory and ordering for all lumber supplies. He is no-nonsense, precise, and work-oriented. It is hard to imagine the store functioning without him. Sean has bachelor degrees in physics, math, and biology, a master's degree in marketing, and a Ph.D. in microbiology. Sean's wife is an equine veterinarian, and they have a house full of adopted animals, mostly dogs.

This is how the economic wheels keep turning in America and around the world—with dependable, embedded IMPACTS energy, dedicated to excellence.

Maria is a Colombian who has been living in the U.S. for over 30 years. For 18 of those years she was the director of a nonprofit organization she founded, which was geared to helping Latino youth in the area with academic issues and cultural assimilation. Recently she closed the organization, feeling that she could actually make a more positive difference in the world by being on the periphery. Maria had begun to feel that she had become a combatant in a battle with the unyielding 'established order' as she constantly struggled for funding from city, state, and other sources. She deemed the fight no longer worth the expenditure of energy; she felt it was having a deleterious impact on the kind of person she wanted to be. Maria has resumed one of her true loves—artistry through painting. Her husband is Caucasian and travels extensively in his job with a major technology company. I mention his ethnicity only to show how pervasive multiculturalism is among the IMPACTS.

Maria's experience illustrates the difficulties of bringing excellent ideas from the periphery into the mainstream. The hurdles are high; the SN of the governmental entity or industry is generally fiercely resistant to change. That is why most good ideas never become incorporated.

Alice is a twenty-year-old with a big dream—to compete in the Olympics as an equestrian. Home-schooled for two years of high school, Alice then attended a community school where she excelled in engineering, drafting, and architectural design. The following year she attended the local university's post-secondary program. But now she is not in school; last year she moved to North Carolina to pursue her dream and is currently being mentored by skilled professionals. Her mother, a bookkeeper, nutritionist, community volunteer, and Far Eastern 'energy' adherent, moved here to offer support. Plus, the ever-resourceful Alice is working two jobs, one at a pet store and the other at a horse farm.

In an interesting twist to the story, Alice's brother works for the CIA. This may sound cognitively dissonant, but actually it makes perfect sense if you understand the IMPACTS concept. IMPACTS are found 'off the beaten path', often in what we might call niche areas. Remember, IMPACTS are filling in the blanks which generally attract little interest from others. Combine that with passionate caring, precision, a commitment to excellence, and protection of strong bonds, and you have a fairly normal IMPACTS family. The brother is working to protect the bonds of the country while the mother has moved down to protect the bonds of the family. Others of course would say that this family is on the periphery, but as we continue to see, that is where the IMPACTS frequently reside.

Sonya came to the U.S. from Trinidad by herself twenty years ago. She has had various jobs, including managing a bagel store. Currently she is cleaning houses, aided by her eighteen-year-old daughter who wants to be a veterinarian or a lawyer. "Really, I just want to help people." A customer of Sonya's services says that Sonya is always problem-solving while performing her job. "This needs to be fixed and I can do it." Sonya is also caring for a foster daughter.

Dora majored in special education and eventually made her way into politics at the local level, serving two terms on the city council during tumultuous pro-growth versus prudent-growth times. Prior to her service on the council, she took command of fund-raising for a local park where kids with and without disabilities could play together. Recently, the park was named in her honor. Dora's husband is a successful entrepreneur in the accounting field, and they have two adopted golden retrievers. Her family lives in different areas of the country but remains very close.

All of those discussed are just 'the-people-next-door-types', but let's look at them a little more closely and see if we can see some patterns. All are mobile, and family is obviously important. Personal initiative is prominent, as are health concerns and people concerns, often with a cutting-edge technological element as would be found in critical-care nursing, veterinary care, and intelligence work. A love of animals is shared; most have dogs, some have cats, and Alice has her own horses.

There is something else that will prove to be very important as we go forward—many IMPACTS will be in roles that have traditionally been held by the opposite sex. As I mentioned, IMPACTS are usually going against the grain or even starting a new grain. Paul was a nurse in the Navy; Susan and Ann are working in a so-called man's world; Alice has her eyes set on succeeding as an equestrian; Sean's wife is a horse doctor. It also was not that long ago that local politics was a man's world almost exclusively.

Two more things: all of the aforementioned are very hands-on in their approach to work and life, and most live in environments buffered by a body of water, a dense stand of trees, open land, or some other natural or man-made barrier. IMPACTS want to be in the middle of the action at work, but they want privacy and space when they get home. When we get to Chapter 4 and the San tribe, we will discover the WHYs of all of these characteristics.

Technology Rebels

The ascendant Linux computer operating system, developed originally by Linus Torvalds in the early 1990s while he was a student at the University of Helsinki in Finland, has grown to rival the Microsoft empire. When looking at innovation, it is important to look at the context from which IMPACTS ideas arise, most importantly the personality profile and family history of the innovator. But location is very important as well, along with other factors such as the political and economic environments. Most of the time, we are going to see peripheral agents intertwined with strong nurturing and a fascination with learning within the family, and an imbalance of some sort. The imbalance may be in the family, the psyche, the community, the economy, the country, the world, or the natural environment. IMPACTS try to correct imbalances, sometimes consciously and always subconsciously.

Bill Gates came from a very nurturing, community-oriented family. He was on the periphery of the nascent computer industry; he was young, he dropped out of college, and he had no experience in the business world. From that context arose Microsoft. Obviously, experience had nothing to do with Gates's conviction that he could succeed, and knowledge of the industry had little to do with his success. It was his drive, his intellect, and his belief in his own abilities that made the difference, and that came primarily from his IMPACTS genealogy and his social environment. Of course we can be certain that he had some lucky breaks along the way also, and he always had strong support from family and friends.

Others would say that in many of his achievements, Gates walked a fine line between legality and illegality, ethics and the opposite. But as noted earlier, today's world is an SN-IMPACTS hybrid, and the business world epitomizes that.

Linus Torvalds, whose parents were both journalists, is actually from a minority group in Finland whose first language is Swedish, not Finnish. There is the peripheral element again, including a minority aspect. You will often find IMPACTS between things, facilitating solutions. Straddling two worlds is very much like the valence electron; it is attached—barely—to the nucleus (structure) and is looking for other possibilities that will bring balance to the situation.

IMPACTS are very often in the same situation and doing the exact same thing. Being 'off-the-beaten-path', the IMPACTS-person subconsciously starts looking for ways to 'fix things'. That is usually accomplished through creative-formative-production of some sort that helps people, helps connect people, and/or helps people solve problems. But it may also be accomplished by starting over in some sense. This can include the formulation of a new way of viewing reality; e.g., the works of Einstein, Newton, and others. Either way, something is going to be invented—the new beginnings element—that will hopefully bring more balance.

Journalism as the career choice of Torvalds's parents is an indication of the IMPACTS orientation of the family. Journalists are also in the middle, between the event and the reading public. They are trying to decipher what actually happened and deliver it objectively. It is not unlike the computer programmer who is taking something that is not understandable and is making it so. IMPACTS, more than most it seems, try to see the picture as it really is. Why? Because it is efficient and therefore aids in the identification of needs. The more accurate the diagnosis of the environmental situation, the better chance of a solution, and that is the ultimate goal.

The complicated ethnicity we see in Mr. Torvalds is very often a part of IMPACTS as we saw with Paul. Though Finland appears to be 'out-of-the-way', its peripheral location is actually a clue to its IMPACTS roots. It resides in the same Baltic Sea neighborhood with Sweden, Norway, Estonia, Denmark, and St. Petersburg, Russia, areas with a reputation for progression and innovation. Finland is the home of Nokia, the cell phone manufacturer, and St. Petersburg is where the Russian Revolution started. What does that have to do with anything? It is just a reminder that innovative genes are rebellious genes—against the status quo.

Peripheral areas such as Scandinavia are havens for IMPACTS genealogy. Other examples are Iceland, Scotland, and New Zealand. Such areas long ago provided a measure of seclusion and sometimes protection from the expansive SN. Groups gravitated to these types of environments so they could start again with 'new beginnings'.

Torvalds began attending the University of Helsinki in 1988 where he studied computer science, and started working on an operating system as more of a hobby than anything else. This operating system, introduced to the world in 1991, is open-source, meaning the code is open for anyone in the world to see and possibly modify. It is not proprietary, and there are no secrets. You can see the borderless-communal-sharing aspects, often very indicative of IMPACTS ideals. The differences with Bill Gates are clear and visible. Bill Gates resides comfortably within the SN, and although Torvalds is gradually making his way to the center from the periphery, he does not appear to be all that comfortable being involved with the SN.

The story of the rise of Linux as a strong competitor to Microsoft is a story that has been repeated in some form or other thousands of times since the emergence of agriculture. The semi-rebel, Bill Gates, became the leader of his industry. But no matter where you find a structure, you will find IMPACTS on the periphery observing and saying to themselves, "I think there is a better way to do this." That thought is as natural to an IMPACTS-person as is breathing, and IMPACTS can become consumed with improving the prevailing structure, or with building a better version of the existing one.

The Linux family is a prototype of IMPACTS creative-formative-productive energy (CFPE). A connecting vehicle for IMPACTS around the globe, Linux is unique in the way that it develops and delivers its product(s). There is Linux Inc., but it is run like no other major corporation in the world today. It has no CEO, no headquarters, and no annual report. It is more of a consortium than anything else, a hybrid between the traditional giants of the technology world—Hewlett-Packard, IBM, Intel, and others—and the iconoclastic Linux open-source crowd. Quite a marriage. Honestly, though, Linux Inc. is not that different from the hydrogen atom in its structure in that there is a creative side and an accretive side. Most structures have the same design and construction.

Linus Torvalds is the defacto head of Linux Inc. but only because he is trusted and has a proven track record. If the trust started to wane, leadership would pass to someone else, not through a vote but through the natural pull to a new, qualified, proven leader. It sounds very organic and very much like the structure of the San tribe of Africa, genetically the most diversified human group on earth and therefore the oldest. We will learn more about the San in Chapter 4.

Financial reward for much of the Linux crowd is not the primary or even secondary concern. It is more about making this particular tool accessible to people around the world than it is about market-share and returns-on-investment (ROI) and all of the rest of the Wall Street mumbo-jumbo. Many of the global workers who contribute just want to be part of a challenge that they believe is valuable and positive. Plus, they enjoy the autonomy and camaraderie that come with the industry. Many just delight in solving technical problems.

IMPACTS are always looking for the most efficient way, which means conservation of energy. Open-source is efficient and relatively inexpensive, and non-exclusive. It feels as much like a community as it does a business.

We will see that the emergence of agriculture about 10,000 years ago is the clear dividing line in modern human development between one worldview and another. Linus Torvalds represents the pre-agricultural mindset and Bill Gates represents the post-agricultural mindset. The earlier mindset obviously did not disappear; it just became 'captured' and embedded.

This brings up an interesting comparison. The peripheral, or P-IMPACTS, generally want to share their innovations with the world with very few strings attached. That is because they generally prefer an economically egalitarian world. The SN-IMPACTS want to share their innovations too—if the price is right. The SN political leaders themselves want the SN-IMPACTS to maximize profit because that means more power and control for the SN leadership.

Of course there are differences among SN leaders, and there are also differences in how countries handle the whole SN apparatus. Some SNs are more people-oriented while others focus more on the leader and the structure. Overall, the present tendencies of the SN in the U.S. are distinctly tilted to the structural side and away from people concerns. This means that the SN is heavily accreting energy from the people in order to build up and protect the SN entity, its allies, its view, and what it stands for—its version of a free market economy and its version of how the U.S. should conduct itself on the world stage. People come and go—the edifice is forever, or so the SN thinks. If the health of individuals was a concern, we would see it clearly. Rather, it is the health of companies that is the major concern. Why? Because companies are seen as contributing more value to the SN than are individuals. The edifice has priority. The people serve it, not the other way around. It demands deference and respect, whether it deserves it or not.

So around the world we will see societies that tend toward a stronger SN and weaker people-orientation, and we will see others that lean more toward people and less to the SN. Actually, what we have worldwide is what we would expect—we have a hybrid between the two.

Red Hat, a company based in Raleigh, N.C., has made a successful business of packaging and servicing Linux products. Bob Young, one of the co-founders and a former CEO, left the company a few years ago and then started the online self-publishing company, Lulu. Authors from around the world have utilized Lulu to self-publish over 750,000 books in the past five years. Lulu also sells artwork, music, software, videos, and more, produced by artists who until recently had few outlets for their work.

Bob Young, originally from Ontario, Canada, is a good example of IMPACTS energy in action. First, he co-founded an open-source software company; then he started a company that tries to help 'regular' people get their creative products out into the world. Recently the company moved into a remodeled former equipment company building with a huge bulldozer on top. There again, he is exhibiting P-IMPACTS traits as he eschews the slick new buildings downtown. What Bob Young seems to be saying with his life's work and his choices is, "There is value on the periphery that is not being utilized. Let's don't leave anything or anyone behind. Let's reach out, get more people involved, and improvement will be the result." That is the basic philosophy of the valence electron as well.

Clearly, Wall Street has had wondrous days and returns because of what the neophyte Gates brought to the business world. But here is a ragtag Linux bunch that has defied the modern world's approach to marketing, distribution, ownership, and the usual ways of conducting business. Now this rebel clan is gradually being merged with the white-collar culture of IBM, HP, Intel, and many others. It is all being done out in the open, but that does not mean that it is all even-steven. It is obvious that the big SN-favored corporate guys will emerge as the financial winners, but then that is not the driving motivation for the open-source crowd. The differences between these two groups are clear, and they are instructive. They are the same two forces we will see at work throughout the book, though they will take different forms. The crux of the matter is that the SN wants to accrete the innovative IMPACTS energy so it can further its goals, and it will usually get what it wants one way or the other. It might be peaceful, or it might not be peaceful.

Legions of followers around the world work on improvements to the Linux system. The Internet enables them to work as a team without a clearly-defined nucleus, which is how IMPACTS usually prefer to work. That is how Torvalds and Linux Inc. operate. Again, we will see the exact same thing with other creative-formative-productive energy, such as bacteria (prokaryotic cells) and the aforementioned group called the San, who also lived in a community structure with an ill-defined nucleus. Such a design allows information and communication to flow much more easily in and out of the structure than does a nucleated structure where roles are more clearly defined.

The Big Corporate Guys have accommodated themselves to this anti-business open-source crowd, but at one time many of them were part of the same category. In 1937, Bill Hewlett and Dave Packard, two electrical engineers with $500 and no product ideas, started the Hewlett-Packard Company. The HP Way "reflected a deeply held set of core values that distinguished the company more than any of its products. These values included technical contribution, respect for the individual, responsibility to the communities in which the company operates, and a deeply held belief that profit is NOT the fundamental goal of a company."[1]

Whenever an employee was discovered to have violated HP's ethical principles for a short-term increase in profits, the individual was fired, no matter the circumstances or the impact on the bottom line. The reputation of the company had to be protected. Many executives from other companies did not think such un-business-like people belonged in the business world.

Many of the establishment companies which are part of Linux Inc. often hire programmers who were formerly the aces on the periphery. Programmers, a very IMPACTS group, are the heart and soul and brains of the Linux phenomenon. Torvalds now lives in the U.S. but makes a fraction of a fraction of what the big corporate guys make. The Open Source Development Labs, those companies with a stake in the success of Linux, pay Torvalds about $200,000 a year. That is cab fare for some of the SN-favored crowd. Torvalds shuns the spotlight and virtually lives in a virtual world. He is still on the periphery even as the SN pulls him in. And make no mistake—that is what is happening and has happened repeatedly since the development of a more settled life after the emergence of agriculture. Torvalds has made his way from Finland to the center of the technological world. The technology capital has not gone to him, evidence of the true dominant force. The reach of the structural-nucleus, the SN, is long and powerful.

Browser

Many have compared Tim Berners-Lee's invention of the worldwide web browser to that of the printing press by Gutenberg. The effect on civilization could be similar. Berners-Lee, from England and a graduate of the Queen's College at Oxford University with a degree in physics (1976), was the child of mathematicians who both worked on one of the first commercial computers in Britain. His father was also enthralled with the workings of the brain. I have mentioned that family history plays an important role in identifying IMPACTS. Each parent worked on cutting-edge technology, his mother worked in a traditional male domain, his father was interested in the most important organ, the brain, and they both possessed that 'objective' element I mentioned earlier—in this case, mathematics. You will find a lot of IMPACTS in exceptional fields but just as many in the regular day-to-day areas performing exceptionally.

A programmer by profession, Berners-Lee was looking for a memory substitute because, according to him, his random-connections quotient was low. While working in Switzerland for CERN, Europe's particle-physics laboratory, he devised the workable idea of connecting information across the Internet and introduced it to the world in 1992. IMPACTS are efficiency-hounds, as we have noted, and are always looking for a better way, which often ends up impacting the entire world just as Gutenberg and Einstein did.

Berners-Lee toyed briefly with the idea of commercializing his invention but instead decided to become a lifelong advocate for optimization of the worldwide web by keeping it open and free from the entanglement of competing interests. Optimization is a basic part of the IMPACTS philosophy.

Here again we have the mining aspect. Berners-Lee has discovered unrealized potential, which he is sharing with the world rather than enriching himself financially. This is the way the

IMPACTS profile started, as we will see with the San—with an unselfish sharing attitude. The SN, through natural selection, has gradually pulled IMPACTS energy toward the middle, producing SN-IMPACTS who generally behave more to the SN's preferences. The peripheral P-IMPACTS usually behave more like the original IMPACTS. This is the way it works with energy fields—the more powerful one pulls in the less powerful, or accretes it.

Berners-Lee's goals for the web browser were typical IMPACTS ideals. It was meant to be a social place where people could work together in a creative and expressive environment, which sounds unsurprisingly like the Linux crowd. Berners-Lee is a member of the Unitarian-Universalist Church, a church that believes in "the inherent dignity of people and in working together to achieve harmony and understanding". IMPACTS are attracted to Unitarian-type churches where the emphasis is on service to the community and its individuals, and on the connectedness of all human beings, not on ceremony and dogma supported by an extensive bureaucracy. Berners-Lee also has an artistic side and plays the piano, both of which we see frequently among IMPACTS. Berners-Lee is obviously a P-IMPACTS-person, more comfortable away from the power-control SN and its emphasis on money and possessions.

We have discussed three prominent IMPACTS figures in today's technological world—Bill Gates, Linus Torvalds, and Tim Berners-Lee. This is a perfect representation of the dynamics that have been at work for thousands of years. Gates comes from a very nurturing family, and as we mentioned, appears to have strong structural-nucleus (SN) proclivities, much like his longtime adversary, Steven Jobs, who we will discuss later. The Linux inventor Torvalds, on the other hand, appears content to live a rather spartan life, especially as it compares with the economic value of his creative-formative-production. But wherever there is value in today's world, a strong, nucleated, hierarchal structure will either form right on top of that value, or an existing SN will snatch it if given the chance. And P-IMPACTS often give the SN the chance because power-control is generally anathema to them. They usually just want to create and produce something that has function and value, and they want to do it with excellence. It was the same with Berners-Lee. He could have gone the commercial route, but he was not comfortable in that arena.

Note also the geographical locations of the three gentlemen we discussed—Scandinavia, England, and Seattle. Here again we have peripheral areas with lots of water. IMPACTS throughout the development of modern humans stayed close to the open sea because it was their lifeline.

Mining

IMPACTS are miners. What is being mined or extracted is unrealized potential. If it can be discovered and pulled out of its 'cave', it can be utilized to help answer needs. Then it has VALUE. IMPACTS have the same attitude no matter where they are in life. Helping a child who has difficulty learning is mining for unrealized potential. So is searching for new supplies of oil and gas or any other natural resource.

Today the Internet is assuming a prominent place in commerce and in people's personal lives. Let's look at one of its mainstays, Google. First, what is Google really? Yes, it is a search engine but at its core it is a mining company—it is mining for information. People are searching the billions of pages that are available online for *nuggets* that they can utilize. All of those billions of pages have no value in the Google world until they are mined and discovered and can be put to practical use. It is just like a goldmine.

Let's look briefly at the founders, Larry Page and Sergey Brin. First, the genetics. Larry's father was a computer science professor at Michigan State and one of the early pioneers in computer science and artificial intelligence. His mother taught computer programming at Michigan State. All of this indicates strong IMPACTS genealogy, especially pioneer, artificial intelligence, and programming. Please note the 'objective' or exacting nature of the professions. Being an early pioneer is akin to being an explorer, a miner really—trying to realize potential that lies undiscovered. Artificial intelligence is another way to be efficient, to get even more brain power than you currently possess, to duplicate. That is why we see so many IMPACTS around robotics.

Sergey Brin's family emigrated from Moscow when he was six years old. His father is a mathematics professor at the University of Maryland and his mother, a mathematician and civil engineer who works with NASA. As a young boy, Sergey attended a Montessori school.

There are many IMPACTS elements here, the first being the emigration factor, leaving for better opportunities. 'Peripheral' Montessori schools are found often in the world of IMPACTS, as are mathematicians and female civil engineers. Why female perhaps more often than male? For a male civil engineer, it is a job that is on the beaten path. For a female, it is on the periphery. And again, the professions have a strong exacting nature.

The Google owners exemplify IMPACTS elements: they drive environmentally-friendly cars, allow roller-hockey games in the parking lot, have an on-site masseuse and a piano, and seem un-enamored with all the money that surrounds them. Their company motto is "don't be evil".

This is typical IMPACTS behavior. Look at the peripheral elements. The parents are all in academia and government work—not business. One of the founders was a recent immigrant. The founders do not appear to be regular businesspeople. The whole thing looks atypical but is, in fact, very typical. We have just never looked closely at the dynamics and processes—and the people—involved in business development. The IMPACTS invent the new which then becomes a part of the structure, changing the structure at the same time. Again, it is the valence electron at work.

We have mostly mentioned men so far but it would be ridiculous to think that women are not changing the world too and have not always done so. Of course they are and they have. It is just that women were 'captured' by the SN in a particular way as it developed alongside agriculture. It would not be too much of a stretch to say that they were enslaved, but so too were all the IMPACTS and everyone else, not to mention nature herself. Most of the energy was accreted by a very few, almost always males. It has only been relatively recent that the chains have been loosened, though not broken.

We are seeing that this is the template for a hierarchal structure. A tiny power-control group at the top controls most of the energy, or certainly attempts to do so.

Bacteria

Bonnie Bassler is a person with abundant IMPACTS energy who is changing long-held assumptions about the natural world. Bassler is a professor of molecular biology at Princeton, but she is not your typical scientist. Every morning she walks a mile down to the local Y and leads a class in aerobics, which she has been teaching for over 20 years. As we have already seen, IMPACTS are often found around health issues.

Bassler loved animals as a child so she decided she wanted to be a veterinarian, until she discovered that she fainted while dissecting little critters. So she volunteered to work in a biochemistry lab and realized that she had a fascination with bacteria. Bassler has discovered that bacteria are extremely communicative, utilizing chemical molecules to 'talk' with their own species and bacteria of other species as well.

Bacteria can behave in effect as a multi-cellular organism when these molecules reach a certain density called quorum sensing. Quorum sensing allows bacteria to coordinate the production of the biofilm formation that you feel on your teeth in the morning, and to release virulence factors when threatened. Bioluminescence (light) is also generated by some bacteria when this quorum is reached, and these bacteria are often accreted by other creatures in the water to facilitate their own survival.

One reason that bacteria can be so prolific and adaptable is that they can exchange DNA with one another, even across species. There is the value again of not having a clearly defined nucleus, which would prevent such communication and exchange. It is the same with humans; smaller groups without a true 'nucleus' can be more adaptable and flexible.

As bacteria laid the foundation for the diverse forms of life that were to follow, communication appears to have been embedded, which helps us understand the nature of human cellular communication. After the importance of Bassler's discovery had sunk in, many scientists were obviously thinking: "Duh—well of course bacterial cells would communicate with each other. Anyone can see that. Why didn't I think of that?"

We do not think of that because we have divorced ourselves from the web of life, creating a special spot for ourselves in nature, being part of it and not part of it at the same time. It appears that our unique consciousness has seduced us into believing that we are somehow exempt from the social strategies in which other animals, and even bacteria, engage. But it turns out that we are in fact no different; we have our own hierarchal structures and social strategies that bear a striking resemblance to those of other creatures, large and small.

Our consciousness should actually make us more aware of what is going on within us and among us, but instead it appears that it allows us to deceive ourselves as much as anything, or to be led into deception and used. That statement is not a swipe at human beings; it just demonstrates the power of the SN to control our collective consciousness, much as a black hole controls the galaxy. We can be certain that human consciousness was very different—much freer—during the long period that preceded the development of agriculture.

Bassler illustrates a couple of other things that we often see in the IMPACTS profile. First, she is not in a traditional role. Until very recently, most academic positions such as Bassler's were held by men. Thus, she can be thought of as a peripheral agent in her particular field. Peripheral agents break the equilibrium of the existing order and open the gates to change; they are pioneers. The reason we have not seen IMPACTS in roles more suited to their talents until the last few decades is because the structure of the human world since agriculture emerged has, for the most part, been opposite from its natural foundation. The pendulum is swinging back, however, because of IMPACTS actions and movements over the past 50 years. More is needed. We will discuss this in detail later.

One more point. Ms. Bassler is a molecular biologist. IMPACTS can be found often in any field with 'bio' in the name, such as biochemistry and biotechnology. Why would that be the case? Because the IMPACTS are part of the creative-formative-productive energy just as bacteria are. Thus, we can assume that they have a subconscious identification with the beginnings of life and biological processes, or certainly a strong appreciation of them.

Oprah

You cannot talk about women changing the world and not mention Oprah Winfrey. Oprah is a good example of how the periphery moves to the center, or maybe I should say of how the center moved toward the periphery in this case.

Oprah Winfrey had a rough beginning in life. Born in Mississippi, her unmarried parents parted shortly after her birth, and she continued to live in Mississippi with her maternal grandmother. There she learned to read at a very young age and was known locally for her exceptional public speaking, before she was five years old. At age six, she moved to Milwaukee to live with her mother, who was a maid. Circumstances were difficult; Oprah was often forced to sleep on the front porch by the woman who owned the house because this woman, a light-skinned African-American, did not care for those with darker skin.

As Oprah grew older, she was abused sexually by a cousin and by other males. As anyone might expect, her behavior became very rebellious. At age 14 she moved again to Nashville to live with her father (it had been tried before), who introduced a strict regimen based on education. He himself had gotten his high school diploma at age 25 and believed strongly that education was the foundation for Oprah's future. She had to read a book a week, combined with a book report, and had to learn five new vocabulary words a day. We all know the rest of the story: the climb to the top of her profession and her work and generosity on behalf of others, including the recently opened school for young girls in South Africa. But we are missing some key IMPACTS elements in her family history and early environments.

First, her mother. IMPACTS come from all walks of life, and many IMPACTS do not get the opportunities and support that they need in order to develop their potential. Her work as a maid may not sound impressive to the person on the street, but in fact we will find many IMPACTS in jobs which are providing a service to others. Plus she had moved to Milwaukee to get a new start; relocating for better opportunities is something IMPACTS often do. Later, Oprah's mother attended classes at a local community college and then worked in the kitchen of a hospital where she became the supervisor until she retired.

There are IMPACTS traits here: the efforts to improve herself is one and the job at the hospital is another. She could have worked in a kitchen anywhere, but she chose a hospital. That is an important sign.

Oprah's father, Vernon Winfrey, was a sailor in the U.S. Navy who later owned his own business as a barber, also entering local politics and serving on the city council. His father had owned his own 250-acre farm when most African-Americans were still victimized through sharecropping. His father had 'stayed in his place' to protect his children, but then aided in the civil rights movement. Vernon's great-grandfather had placed a school for blacks on his property after the Civil War when African-Americans were struggling to obtain an education of any sort.[2]

Vernon's discipline was intended to help Oprah be the best she could be. It appears that it succeeded to a large degree. IMPACTS come in all colors from all professions, from the country and the city, from the north and the south, from every country and every nook and cranny on earth.

There were IMPACTS traits in Oprah from the beginning. This in no way deemphasizes her Herculean accomplishments. It does however point out the fact that there are many gifted IMPACTS out there who never have the opportunities to move forward because the structure that is in place neither sufficiently supports nor allows for the development of their abilities. The barriers can be insurmountable, even for the very talented and motivated.

The SN generally cares little for those on the periphery because the SN sees little value coming from that far away. Keep in mind that in the atom it is the valence electron which goes out and bonds with other peripheral electron energy. It is the same in the human world; the valence electron IMPACTS are usually the ones who go out and try to take care of the others who have been left behind. They try to prevent dispersal by bringing everyone inside the group.

You can see many IMPACTS elements in Oprah's story:

- A loving grandmother who encouraged her to develop reading and speaking talents at an early age.
- Her mother's occupations involving service to others along with her self-improvement efforts.
- Her father's entrepreneurialism, his service to the community, his service in the Navy, his focus on learning and education, and his attitude that Oprah was expected to be her best possible self.
- The periphery factor—Oprah was nowhere close to the mainstream when her life began.
- Her generosity and her desire to make a difference in the lives of others.
- The book club she started in order to encourage reading and authorship.
- Her gifts to South African girls and their education. For the young girls, change is also coming from the periphery just as Oprah is bringing change to America from the periphery.
- The cross-cultural element—America and South Africa. To most IMPACTS, the world has no borders.
- Oprah's rebellious streak. Of course some of it was related to the sexual abuse, but rebelliousness is often seen within the IMPACTS, even beyond the normal teenage examples. Rebellion is dissatisfaction with the status quo, and that dissatisfaction is the foundation of the IMPACTS profile.

The human genome, in my opinion, has a savior 'gene' embedded due to the beginnings of the IMPACTS profile and its importance to human survival, which we will explore in Chapter 4 with the San. Not all IMPACTS are saviors, but occasionally one comes along who merits the title.

Oprah and Tiger Woods (below) represent how an IMPACTS-person can come onto the scene—from the periphery—and change the standards of the profession, or even society, forever. The reason is that the middle becomes very homogeneous and resistant to change. Therefore, it requires considerable departure from the norm to puncture the equilibrium. Peripheral agents less talented or formidable would not be able to break through. That is why it requires an Oprah or Tiger or Martin Luther King, Jr. to knock down the walls.

Tiger Woods

I have mentioned that we often find IMPACTS with multiple cultural lineages. Tiger's father, Earl, was a mixture of African-American, Chinese, and Native American ancestry. Wood's mother, Kultida, is originally from Thailand and is Thai, Chinese, and Dutch. There is a lot of emigration and immigration in the ancestry of Tiger Woods.

Tiger has stated that he wants to be better known for what he does for others than for his golf. That will be a tall order to fill. But his efforts to help youth maximize their potential, including the Tiger Woods Learning Center in southern California, are serious attempts to affect IMPACTS concerns of youth and education.

Everyone is familiar with the intensity of Tiger Woods and of his quest for excellence. After winning the Masters by 12 strokes in 1997 and the U.S. Open by 15 strokes in 2000, Tiger decided to change his golf swing. IMPACTS are rarely satisfied with themselves. It is not a negative feeling. They just want to be better. Actually, they want to be their best because they were born with that attitude.

If we look at golf a little closer, we will see further examples of the peripheral agents coming in and influencing the existing structure and culture. American women's golf has become international women's golf with stars like Annika Sorenstam from Sweden (recently retired), Lorena Ochoa from Mexico, and a host of South Korean stars. Michelle Wie is an American golfer whose parents emigrated from Korea. Her family has a strong background in education, a focus seen frequently among IMPACTS.

The influence on culture is also true of men's golf and every field imaginable as IMPACTS immigrants flow into the U.S. from every corner of the globe. But no longer does the U.S. have a near monopoly on economic opportunity. Now, there are many options around the world for those who are seeking greener pastures.

IMPACTS are looking for places where they can 'realize potential', help their loved ones, and contribute to humanity, and they will get on a plane or bus or walk if they have to—if no one stops them. After they arrive, they will exert a powerful impact on the social, cultural, and economic environments. This kind of movement has been a signature of IMPACTS and their ancestors for over 100,000 years.

Tiger's father, Earl, was a career military man in the U.S. Army, retiring as a lieutenant colonel after 20 years of service. He also served two tours of duty in Vietnam. Earl's father died when he was 11 and his mother died when he was 15. He was raised by one of his older sisters for the next few years. Earl received a baseball scholarship to Kansas State University where he broke the color barrier in the Big Seven Conference in 1951. He graduated from college and was offered a contract to play in the old Negro Leagues, but instead started a career in the Army.

It is easy to see where Tiger gets his competitiveness and his patriotism. Some have suggested that Tiger should be a stronger agent for social change. But at this particular time, he is comfortable working at being the greatest golfer of all time, raising a family, and giving what he can to young people and military people, especially those injured in battle. IMPACTS try to make the world better in their own special, unique ways. Tiger adds tremendously to the world just by being who he is—a striver of excellence and a person who cares.

Tiger has put most of his 'dissatisfaction with the status quo' into his golf game. If he had put any of it into the political arena in a controversial area such as expressing an opinion about U.S. policy in the Middle East, he would probably have enjoyed much less acclaim. The SN expects you to keep your talents confined to your field, unless you are prepared to help the SN. The price can be high for going against it. The SN 'naturally selects' who will get through its filters. Currently, Tiger shows all signs of being trustworthy as far as the SN is concerned. Consequently, he has many lucrative advertising contracts with car companies, golf equipment manufacturers, and others. If he 'steps out of line', he will immediately be kicked to the periphery and will forfeit all SN support.

Delivery

Did you know that Walt Disney and Ray Kroc, founder of the McDonald's franchise system, both volunteered for the same kind of job with the same organization when they were teenagers? The position was an ambulance driver with the American Red Cross. With most IMPACTS, the welfare of human beings is the number-one concern. It is no accident that both went on to deliver other forms of value. Disney provided family entertainment and enjoyment for decades, most of it focused on children, and Kroc was really the founder of the modern franchise system. IMPACTS want to find ways to duplicate value, to extend it as far and wide as possible, over and over again. Film is one way to do it; franchising is another.

Kroc discovered the McDonald's restaurant, owned by the McDonald brothers, because of one thing—he was selling them milkshake machines, and they were putting them to work. For seventeen years, Kroc had been crisscrossing the country selling his machines. But there was something unique about the ones he sold, which reveals a clue as to how he saw the world. You could make eight shakes at one time with his model rather than the single shake produced by the conventional ones. That is what attracted him to this particular product—the duplication-efficiency factor. That is what he saw when he observed the original McDonald's restaurant—the potential for duplication and replication of value. So Kroc paid them a couple of million dollars for the rights, and he was off to the races.

Kroc did not start the McDonald's franchising business until he was in his early fifties. IMPACTS do not think it is ever too late to add value. Of course, some people will say that he created a disaster. Either way, it was a formidable achievement by a very regular guy—from the periphery. Kroc was not part of the corporate elite; he was a traveling salesman. By the way, he was also a pianist. It is amazing how many pianists we see among the IMPACTS, far more than any other musical instrument.

I mentioned Einstein earlier. Einstein was an employee in the Swiss patent office and basically separated from other scientists and theoreticians of his day when in 1905 in his spare time, he wrote five ground-breaking papers that forever changed our views of the universe. One of those papers would have been a lifetime achievement. He wrote five in one year! And he did it clearly from the periphery with very few connections to the prevailing scientific structure.

What is it about Einstein, Buffett, Gates, Bassler, Kroc, Oprah, Tiger, Torvalds, Berners-Lee, and other IMPACTS that enable them to drive such important economic, social, cultural, and scientific advances? And what about the vast majority of IMPACTS like Paul and Susan who toil quietly and dependably in everyday commerce and society to keep the engines of human endeavors churning? We will see as we proceed that it is not a mystery, that there are clear, understandable, and identifiable reasons. And that it has been so for over 100,000 years.

If you look closely, you can see this one very important element common among the IMPACTS we have mentioned in the book thus far: they are (were) all motivated to DELIVER critically important results, generally in a very hands-on manner. This is a key point to understanding who the IMPACTS are and what they are all about. They create value, they deliver value, and they understand value, though they may not be conscious of the term. They also strive to institute this value in practical ways and to make it accessible to as many people as possible. They want to make the world better every day, yes through knowledge and through the introduction of quality products and services, but also through the development of strong, lasting bonds that will aid humans and also protect them.

The IMPACTS are the ones who go the extra mile when most everyone else would have quit. They have to see it all the way through and know that it has been done correctly. That is why they are in vulnerable situations such as critical-care nursing. They also usually have a highly developed conscience and step forward to help when others may not. The term 'heart and soul' had to be invented with them in mind because that is what they generally put into everything.

IMPACTS are usually family-centered. If they belong to a church, they often serve on committees and/or sing in the choir. They travel to see and experience the world. They try to protect the environment and push for parks and access to nature. You will find them often on and around the water. They frequently adopt rescued animals and help rehabilitate injured wildlife. IMPACTS are the readers and the watchers of informational programming.

They are the ones who go back to school to further their education and dreams. They lead nonprofits and help with fundraising. They serve with Hospice. They tutor. They reach out to those less fortunate through hands-on aid and contributions. They are research scientists, and applied scientists. In the military, IMPACTS serve most often in the Navy. They often adopt children. They promote fitness. They are the social glue and the foundation of human society, everywhere in the world. They are the dispersal hinderers.

All Kinds of IMPACTS

As a group, IMPACTS are not homogeneous. They exist, like everything else in the world, on a continuum, some more interested in creative-formative-production and others more concerned with human health and social well-being. Some IMPACTS are more skilled than others, some are more people-oriented, and some are more rebellious. Many are comfortable being closer-in to the structural-nucleus, the SN, while others prefer life on the periphery. The manifestation of dissatisfaction with the status quo can take many forms. Society works, even if it sometimes doesn't work particularly well, because the IMPACTS are present throughout.

No matter where IMPACTS are, some are more action-oriented and some are more contemplative, though the contemplative ones will still want to put their thoughts and ideas into action. As was noted with Bonnie Bassler and others, gender roles have become increasingly blurred in the economy and society. This is a sign of increasing IMPACTS influence and proliferation and a major contributing factor to progress and production seen at all levels of human society. Still, the walls are high around the political and economic core, the real SN power-control center.

When politicians, sociologists, and historians talk about a strong middle class, they are really talking about the group that forms around an IMPACTS core that is focused on the DELIVERY of valuable goods, services, and information. Think of an economy as a network of IMPACTS nodes of innovative energy supporting a great number of people—similar to a queen bee and the hive. Most businesses and organizations are

IMPACTS nodes, each with varying degrees of IMPACTS innovative and productive energy. But as mentioned, today's world is a hybrid of the SN and the IMPACTS, and most IMPACTS nodes are captured in service to the SN.

I saw this description recently in reference to a nursing association. They were espousing that they were the epitome of the 'Head, Heart, and Hands' phrase. I can think of no better phrase to describe the attitude and actions of the IMPACTS, and how very appropriate that I saw it in reference to a nursing group. Critical-care nurses epitomize the IMPACTS profile more than any other group—selfless, innovative, skillful, alert, available, tireless, dedicated, nurturing, and motivated.

While waiting to pay my respects recently to a friend who had lost his brother in a terrible accident, I asked the lady seated beside me what she did for a living.

"I work as a nurse in ICU."

I asked her if she enjoyed her work.

With obvious feeling she replied, "Yes, especially when I see my patients getting better—much better."

IMPACTS have a passion to deliver positive results, to make the situation as whole as possible, to restore balance. You can see why the SN wants their energy.

Based on Economics

As anyone can see, our world is based on economics. Some might ask, "What else is there?" There is the health of people, the health of the environment, peace, equality, and justice, just to name a few. But these are all IMPACTS concerns, as we will see, and since the IMPACTS are not in control, these issues take a back seat to SN priorities, which are power and control. What is the best way to attain power and control in today's world? It is to dominate resources and markets, so therefore the IMPACTS and their energy have been utilized to a significant degree in the economic arena. Bear in mind that none of this is conscious; it is just the way that 'energy gets organized'.

Theoretically, business is about solving problems, an IMPACTS concern, but today it is more often about making money than anything else. The basic SN philosophy in the U.S. is that you should be the ultimate consumer, focusing on yourself and those very close to you, thus 'dispersing' your energies away from the group. That philosophy is part of the 'break cohesion and bonds' strategy of hierarchal structures. But the recent economic collapse shows once again that the farther one strays from foundational IMPACTS principles, the more risk is incurred. IMPACTS hold everything together but if they are not represented in key decision-making and oversight areas, or if they have been totally captured in service to the SN, results are unpredictable, whether in the economy, foreign relations, the environment, or anything else.

I alluded to it earlier: a vivid example of SN policies that push people apart has been the healthcare situation in America. Most countries in the developed world have ceased treating healthcare as a privilege and instead see it as a right. Countries have learned that if they aspire to reach their potential, their citizens will have to be as healthy as possible. After all, everything flows from the energy of human beings. Why not make their health the primary concern?

Up until now, U.S. legislators have considered healthcare to be a privilege; in this country you have always had a better chance of quality care if you followed the societal template. Currently, the provision of healthcare is primarily the responsibility of employers. So if you can find a 'quality' employer, then you can probably get quality insurance, or so the thinking goes. This pressure on employers adds another burden onto the considerable challenges they already face in the marketplace.

The lack of a thorough, affordable health plan for all Americans leads to unnecessary anxiety, real suffering, and significant numbers of avoidable deaths. It also severely limits the realization of potential of individuals and the society at large. The health of insurance companies has always taken center stage rather than the health of U.S. citizens. As of this writing, even the proposed reforms have an accent on universal insurance but not on universal care. President Obama wants a so-called public option, a plan offered by the government that would theoretically keep the insurance companies in line and make their policies more affordable, and over 60% of Americans are in favor of such a plan. But as we go to press, it appears very iffy as to whether there will be such an option. Will our representatives listen to what the people want or will they pay attention to the desires of those who prop up the edifice? Or will it just be a game of back room deal-making with little attention paid to the wants and needs of the citizens?

Prior to this attempt at reform, the SN has been 'naturally selecting' those that it believes will be the least drag on the system. It has been saying, "If you follow our preferred approach, you have a good chance of 'success'." It has been fostering and supporting a class system—a homogenized center and a disenfranchised periphery. But that is what a hierarchal system does; it pulls in the energy that will help it get bigger and stronger and discards the rest.

Unfortunately, politics is leading the discussion which is a recipe for continued disaster. A group of eighth-graders could do a much better job of developing a fair and equitable plan. But that is not what the SN is looking for; it is looking for the maintenance of the status quo with money continuing to flow to the top. Amelioration of human suffering and the realization of potential for each human being are IMPACTS concerns that long ago took a back seat to SN priorities.

What is the SN in this example? It is the insurance industry. The SN is the entity that is in charge, the one with the power and control, and money. The government is actually not in charge of healthcare except in Medicare and Medicaid and other government programs. It is the SN of those programs. The public option is seen by many as the only way to bring balance to the system. It is astounding when you think about it that an insurance company can and does make life and death decisions for millions of people everyday instead of the doctor.

I met a person recently who believed that the family had 'good' health insurance when his wife developed breast cancer. After chemotherapy, she needed several shots that cost $6,000 each. The insurance company refused payment. What would have happened if the family had not been able to find the funds? She would probably have died. And very few would have blamed the U.S. government. They would have blamed the insurance company and maybe the family for not having the 'correct' insurance. But the fault lies with the government and its supporters who perpetuate class-system policies and politics.

A lazy sap who is born into money can get all the quality healthcare needed while the hardworking parents of a family of limited means may have to forgo needed treatment for their children and themselves. Social Darwinism still thrives in the U.S. People can get sick and become destitute through absolutely no fault of their own, and they are often treated as failures and losers. In effect, they are then cast to the periphery. This works to dissolve the bonds that they have developed with others and their community. But this is the 2nd Law at work, manifested in the SN and its policies. It is a carryover from earlier days when peasants and slaves were treated no differently than farm animals, their energy accreted with little regard to their health and well-being.

In many parts of the world, denying a person medical treatment would be considered criminal behavior for neglecting the needs of the sick and injured. Here it is treated as more of a financial matter than a medical matter. But as noted earlier, the U.S. sees the economic engine as its future instead of its people. Its focus is on the production of its citizens rather than on their health. Many lawmakers will vote to ultimately spend more than a trillion dollars on the destruction of Iraq but do not want hardworking Americans to be free of healthcare financial worries. You can see how both of these attitudes—war and limited healthcare options—are examples of dispersal and anti-bonding.

In modern society, the bonds among people are tenuous. Our prisons are filled with men and women, mostly men, who are also quite frequently victims of inadequate bonding. Human bonding civilizes us and helps keep us civilized. Accessible, affordable healthcare for all is probably the best and easiest way to strengthen human bonding, and therefore prevent or mitigate the social deterioration of millions of people who may already have seen other important bonds severed. Plus, it is obviously the right thing to do. That it is even an issue shows just how far the SN has strayed from IMPACTS principles and elementary standards of humaneness.

Karl Marx believed that the story of civilization was the struggle between classes, the bourgeoisie and the proletariat, and presented communism as an attempt to rectify the imbalance. The class struggle that we see is actually the manifestation of the struggle between the two major forces in the universe, the 2nd Law and its violators.

Never Underestimate IMPACTS

Most of the major discoveries throughout history and prehistory have been made by people who were 'off the beaten path'—from the periphery. As we have noted, the leaders of the structure, the SN, try to draw the talented IMPACTS into their orbit. Then the SN leaders decree the societal map that others should follow, defining the terms for success. But IMPACTS, especially P-IMPACTS, possess a different energy than do those in the SN, and therefore are not easily captured, as we saw demonstrated by Linus Torvalds and Tim Berners-Lee. But their innovation usually is.

Truth and creativity more often than not come from the bottom-up. That has always been the case. That is certainly how the development of modern humans proceeded, just as did life itself. It is really only relatively recent, the last 6,000-10,000 years or so, that a strong hierarchal structure has emerged and taken control. Before the development of agriculture, the circle was usually the main structure of society and there was no top and no bottom. Power was shared by all, and issues were generally resolved peacefully, through a consensus. Individual opinions were valued and their expression encouraged. It was real democracy at work, opposite in most ways from what we call democracy today. The SN has even captured that.

In the next chapter, we will look at the theory in more detail.

Summary

IMPACTS are going to be attracted like a magnet to critical positions where responsibility, skill, and decision-making are needed; where people, animals, the environment, or any other valued entity may be in danger; where delivery of valuable goods, services, and information is crucial; and where they can innovate, create, and produce. You might think of IMPACTS as heading for the point of vulnerability in the hospital, school, organization, company, or structure, either to take care of individuals in need or to position themselves at a critical spot where the health of the organization might be affected, such as in accounting. They are on the lookout for problems, which they will try to solve, or potential problems which they will try to prevent from occurring in the first place.

The natural world of the IMPACTS is circular, not hierarchal. IMPACTS are community-oriented and pull for the underdog but they are not monolithic. Some live in the mainstream, others on the periphery. Perhaps the following phrase best describes what the IMPACTS are all about: Passionate and careful delivery of valuable goods, services, and information, deemed to be of a critical nature, to people, organizations, businesses, or anything of value. IMPACTS are found around the cutting-edge because that is where creative-formative-productive energy is. And that is what the IMPACTS are all about—new beginnings.

Usually, where you find IMPACTS, you will find a peripheral element, imbalance of some sort, an exacting nature, and strong nurturing or caring. Change generally comes from the periphery, but the periphery is nuanced. Every structure has a periphery.

The SN tries to bring in all available IMPACTS energy. I call those IMPACTS who reside closer to the SN the SN-IMPACTS, and those farther out I refer to as P-IMPACTS, or peripheral IMPACTS. An SN cannot function without significant numbers of SN-IMPACTS. It is not possible. P-IMPACTS try to keep the SN honest and work as a brake on SN excesses, but the SN generally ignores them.

Footnotes

[1] Jim Collins, Jerry Porras, *Built to Last*.
[2] Henry Louis Gates, Jr., African American Lives, DVD, 2006.

Chapter 3
Theory

Protons do not appear to be as friendly as electrons in that they (protons) seem to have a fierce 'resolve' to stick together. Protons and neutrons are both made up of 3 quarks, which are believed to be the most fundamental baryonic particles. Baryons are particles of matter affected by the strong nuclear force which holds everything together in protons, neutrons, and the nucleus of the atom. The strong nuclear force is the name of the force, a description and a name.

Leptons, of which electrons are the most prevalent by far, are the other major form of matter. Leptons are not affected by the strong nuclear force but they do appear to be affected by the 'massiveness' of the baryonic structure, according to Einstein's general theory of relativity. This again relates to our discussion of the IMPACTS. IMPACTS have an independent streak no matter where they are located in society, but the peripheral P-IMPACTS are less affected by the pull of the SN force. It is the valence electron situation again.

Why do the mysteries of the universe feel so unsolvable, so complicated, so overwhelming when everywhere we look we see the same basic design, a structure revolving around a much smaller and sometimes massive nucleus? We see it in atoms, in galaxies, in modern businesses and organizations, and in the makeup of societies where a tiny core governs the people whether cities, states, or countries. Look at a professional football team. It usually has one or two owners, hundreds of employees, and then tens of thousands of loyal fans supplying copious amounts of energy that help drive the team forward. It all happens around a tiny core. It is the same with a hurricane.

We will also see the same template in the development of human civilization, first as an ill-defined nucleus and then as a clearly-defined nucleus. Shouldn't the ubiquity of the model give us important clues about the overall workings of nature and the universe?

For most of the time before the emergence of agriculture about 10,000 years ago, it appears that people lived mostly in bands of 20 to 50 people as hunter-gatherers, pursuing a harmonious relationship with nature and each other. This was the time of the ill-defined nucleus, which presents a very important point: creative-formative-productive energy (CFPE) usually exists in 'flexible' structures at a distance from the nucleus where it has room to operate, or where there is practically no structure or nucleus at all. That is why we find IMPACTS involved in so many situations and organizations that may appear disorganized or without structure. CFPE by its very nature does not need much support. If it did, it would not be 'new beginnings' energy.

Let me give some examples of 'disorganized' or structure-less situations where we might find IMPACTS:

- Immigration—entering a situation that has many uncertainties
- Entrepreneurialism—the sometimes chaotic process of potentially creating and forming a business structure
- Teachers and administrators at a charter school—trying to build a structure that will help those who are being left behind
- Special education teachers—dealing with situations that are not clearly defined or predictable
- Home schooling—avoiding the typical hierarchal world of education and maintaining the creative-formative-productive-protective circular environment of the family
- The emergency room of a hospital
- The aftermath of a natural or man-made disaster
- Sculptor, painter, potter, composer, songwriter, playwright—an artist
- A developmental editor
- Research scientist
- Inventor.

All of the situations mentioned above operate with little structure; they are all in a way making it up as they go along; they are all trying to achieve the best possible outcome; and they are all on the periphery of the predominant structure. Most are at 'beginnings' or 'new beginnings', including the emergency room and disaster areas, which can be thought of as beginnings in recovery.

The IMPACTS are the bacteria of the human world; bacteria were early creative-formative-productive energy. As noted, bacteria can exchange DNA with each other even across species, enabling them to adapt to new conditions and circumstances. It was probably one bacterium which learned how to make its own food using energy from the sun, and then the DNA got passed around quickly. The storage of energy (food) shows a planning ability of sorts—the extraction of the raw materials required to make food shows a mining 'mentality'.

This kind of rapid transmission of information among bacteria has its precedence in the physical world of subatomic particles with a phenomenon known as entanglement. Though it appears that no actual information is exchanged between these particles, what do we know? Our definitions of 'information' and 'exchange' are very limited. The entanglement phenomenon is such that when subatomic particles interact and then go 'their separate ways', they behave as one, even at vast distances, theoretically at any distance, even the other side of the universe. One seems to know what the other one is doing. Experiments have proven this many times. In a recent experiment, five electrons became entangled, meaning that all five were behaving as basically the same electron or as parts of the same electron.

So entanglement could actually be the ultimate progenitor of this phenomenon of rapid proliferation of information that we see among bacteria. Remember, the living world appears to operate on the same model as does the physical world but at a much slower pace. We will see the same entanglement phenomenon among the IMPACTS also. And we will learn that it started over 100,000 years ago with the San tribe and the San-shaman.

Forces

Modern humans have been around for well over 100,000 years, but the feeling of special-ness appears to be a late addition to our perception of ourselves. The development of agriculture allowed the emergence of a structural-nucleus (SN), which attracted and continues to attract some pretty unsavory characters, many of whom have depicted themselves as chosen and uniquely gifted. Some of this no doubt has often been 'acting', but we can be certain that much of it has been pathological as well, and continues to be. The surplus energy created by the IMPACTS since agriculture has enabled some people who couldn't do anything else to call themselves leaders and actually get away with it.

So the so-called leaders probably introduced the special-ness attitude, and it grew from there. Whoever controls the paradigm generally sets the tone. Post-agricultural leaders were not the same kind of people as pre-agricultural leaders, as we will see.

The philosophy of life, if you want to call it that, of early modern humans before agriculture was an evolving one. Rigidity and dogma had no place in a world that demanded flexibility. As agriculture advanced along with a developing hierarchal structure and an SN core, rigidity and dogma advanced with it. Flexibility became the odd man out. But all of this is really the same battle between the two forces that exist in the atom and throughout the universe, one flexible and the other rigid. A nucleated structure breaks up the cohesive, concentrated creative-formative-productive energy (CFPE) within its sphere of influence and redirects it in ways that enhance the nucleus. The nucleus is a dispersal agent.

To emphasize the dispersal nature of nucleated structures, astronomers believe that the nucleated galaxies we currently see will gradually, over many billions or perhaps trillions of years, become a sea of black holes, which will be just another stage toward total dissolution of all matter. So nucleated structures appear to be the next step after creative-formative-productive activity, but not the last one. As I have noted, I believe that a nucleated structure is a hybrid, or a compromise, of the two major energies, dispersal and anti-dispersal. We will talk more about this later.

In the hydrogen atom, the proton nucleus is mitigating the energy of the electron. Plus the electron has been 'stolen away' from the company of other electrons. Its attitude however appears to be, "How can I make the best of the situation such that the best possible outcome is achieved?" The answer is to find a valence electron of another atom and bond with its energy.

We mentioned the different electron energy levels. The first energy level holds two electrons. That is why the electron in the hydrogen atom is a valence electron—it wants to be paired. The next energy level can hold eight electrons. If it has only four as does carbon, it wants four more, and in the case of carbon, it likes nothing better than to bond with other carbon atoms when possible. When carbon atoms bond with each other and with other atoms such as hydrogen, oxygen, nitrogen, and phosphorous, some of the complex molecules necessary for life are formed, such as sugars, proteins, and DNA. These are examples of covalent bonds, the strongest type of chemical bond. So it appears if you can get a few carbon atoms along with a few other specific atoms, and they all get tied together with their valence electrons, you could have the makings of life. None of it can happen without those searching, connecting valence electrons.

Carbon makes a particularly good atom for building these complex molecules of life because it has 4 electrons in its outer shell and therefore needs 4 more in order to be balanced. Hence, it is stable and reactive, a nice condition on which to build. In that way it exemplifies the dynamic that we said started with the valence electron of the hydrogen atom, passing all the way through life to us. Things happen when there is stability mixed with reactivity. Many of the IMPACTS exemplify the same 'condition' which, as we will see, they inherited principally from the San-shaman.

Atoms can also form ionic bonds and metallic bonds. Ionic bonds form when a metal bonds with a non-metal. Generally, the difference in numbers of valence electrons is high; e.g., in sodium chloride. Sodium has one valence electron in its outer shell and chlorine has seven. This ionic bond produces table salt. Metallic bonding found in metals is another form of electron sharing where the electrons are shared with all atoms in the material. Hence, they are good conductors of electricity. The important point is that everything is happening because electrons are bonding with other electrons.

Looking closely at the atom will help us see the forces at work in human society. We noted that neutrons are an amalgam of proton energy and electron energy, which gives us a clue to the functioning of society. There is some electron IMPACTS energy within the nucleus (SN) of society too, but not much. A neutron outside the nucleus decays rapidly—in about 15 minutes—and changes into a proton, an electron, and an electron antineutrino, a particle that is also part of the lepton family. With the neutron's rapid decay outside the nucleus, it appears that electron energy does not have a natural home within a nucleus unless it is part of a neutron; in other words, unless it is part of structural energy. It appears to be very much the same in the human world. Human structures want creative-formative-productive IMPACTS energy to be tightly controlled because 'running-free' innovative energy can lead to unpredictable results, and an SN hates unpredictability.

In addition to the strong nuclear force within the atom, there is also the electromagnetic force, which comes in a distant second in strength to the strong nuclear force. The EM force exists between charged particles such as a proton and electron, or a proton and proton, or an electron and electron. A force has to have a force-carrier particle, and for the strong nuclear force, this particle is appropriately named a gluon, and it is charged just as the nucleus is charged. The electromagnetic force is carried by the photon, which has no charge. This is instructive also as it concerns human society. Tremendous thick-as-glue-thinking exists as you get closer to the SN of society, but as you move toward the periphery, you usually find more openness and lightness and independent thinking—and less of a charged atmosphere.

Within the nucleus is also found the weak force, but its effect on the universe is anything but weak. Without it we would have no stars and no sun, as it allows for the transformation of neutrons into protons and vice-versa, and for fusion between nuclei which powers the stars. So the weak force aids in transforming the nucleus so energy can be produced. It is believed to be part of the electromagnetic force, and together they are often referred to as the electroweak force. This does not surprise me—that the electromagnetic force and the weak force seem to be working together, possibly against the strong nuclear force.

The fourth known force is gravity, but it is very weak within the atom. Some scientists are still looking for gravitons, the theoretical force-carriers of gravity. But Einstein's general theory of relativity explained gravity in another way. It was not really a force but rather the distortion of space-time by mass and energy. Not everyone got the memo or believed it.

The basic forces inside the atom are operating in human society too. It is the exact same model. We cannot see it because the SN that developed after agriculture constructed a divide between human beings and nature, mostly thwarting open inquiry. The SN distorted space and time around it, capturing most of the creative-formative-productive energy, or IMPACTS energy, which also happens to be the energy of inquiry. The IMPACTS are the force-carrier particles for the energy that bears their name.

Many Einsteins

If Einstein could not figure out the mysteries of the universe, what chance does the guy on the street have? Einstein was brilliant, yes, but certainly millions like him have been born in the last 100,000 years. All we have to do is look at Stonehenge, the 'healing' center in England constructed over 4,000 years ago and aligned with the winter solstice; or the pyramids in Egypt, perfectly situated to north, south, east, and west; or the Parthenon in Greece which was constructed with the use of mega-cranes; or individuals such as the Greek Archimedes who invented integral calculus and mathematical physics in the third century BCE. The examples are almost endless.

Most of the Einstein-like people have made their contributions in obscurity, one reason being that they did not care whether they got any attention or not, and another being that they were too far out on the periphery for anyone to notice. Some of course never had a chance to contribute their talents to the human race. As we are seeing, the very talented and the very rebellious are often housed within the same person, and rebels can have a relatively short lifespan, especially so since the onset of agriculture and a controlling SN. The SN tries to destroy energy that is antagonistic to it. That is what wars are all about.

The SN will generally reject anti-status quo energy unless it can be turned into 'enhancement' energy, which it can be in the

93

economic, scientific, and engineering arenas in particular. The energy that the SN does not want or use settles to the periphery. That is why we see the disabled and other challenging human conditions on the periphery of society—the SN is not being enhanced. Therefore, they are kicked to the curb.

Archimedes, the brilliant mathematician, was not on the periphery of society in the third century BCE in the Greek city-state of Syracuse, on the island of Sicily. His father Phidias was an astronomer and an aristocrat. So Archimedes already resided close to the nucleus; he was an SN-IMPACTS-person and a friend of the king of Syracuse, King Hieron II. When the king asked Archimedes to put his inventive ability to work on practical weapons that could be used to defend the city against the Romans, Archimedes complied, though he took no satisfaction in doing so. He really just wanted to be a pure mathematician, to acquire knowledge for its own sake—to solve problems. Engineering weapons was not his thing.

Doubtless, there were thousands more in his time much like Archimedes, though maybe not his equal, many of them gathered in Alexandria, Egypt. The SNs of course existed BECAUSE of the efforts of IMPACTS such as Archimedes, and the continuing SN challenge would be to get as many of them on its side as possible. They were needed not only for war and weaponry design and construction, but also for civil engineering projects such as canals, aqueducts, and roads. Plus, the personal needs of rulers could get pretty extravagant as well—tombs, pyramids, statues, gardens, and palaces, just to mention a few.

It is the story of the modern world too. The more powerful SNs in today's global political arena try to draw out the kinds of intelligence that will maximize their power and control, and presently that is mostly economic, scientific, and military intelligence. In today's world, economic success gives a country a better chance of attaining military security, but there are certainly no guarantees.

You will notice where Archimedes was located on the societal map of the time—at a very critical point (weapons inventor) and closely aligned with the king. That proximity is where the SN wants all of the talented IMPACTS to be—close, but not too close.

Energy

Einstein proved that everything, including matter, is energy. There is also something called dark matter which is actually about six times more abundant in the universe than conventional matter but nobody really knows much about it. When I refer to matter, it will be the conventional kind.

Conventional matter is the stuff we see—stars, the sun and moon, planets, clouds, the earth and everything on it. It is made up mostly of protons, neutrons, and electrons, what we generally think of as components of atoms. But most matter in the universe is not atomic in structure. It is in a state called plasma where the protons and electrons are separated.

Plasma makes up 95-99% of matter in the universe, or so it is believed. Even the upper atmosphere of the earth, the ionosphere, consists of plasma as the sun's energy, in the form of electromagnetic radiation, knocks electrons off of atoms. Atoms are an aberration in the universe though they are prevalent on earth along with molecules formed by electron sharing.

As mentioned, atoms did not form until about 380,000 years after the Big Bang, when the universe cooled enough to permit it. Plasma is made up of separated protons and electrons because it is too hot for them to form into atoms. For that reason, there are no atoms in our sun or other stars, none in lightning on earth or anywhere else, and none in fire.

Stars form in clusters in cold molecular hydrogen gas clouds, the formation usually initiated it appears through turbulence, such as from a supernova. Without the valence electron forming molecules of hydrogen gas, there would be no stars. The dissatisfaction of the valence electron with its lonely predicament—its imbalance—seems to be the starting point for much of what we see around us.

Does this mean that the good stuff in the universe only happens when atoms and molecules are formed? I do not know. I guess it depends on what good stuff is. But good stuff does seem to occur often when forces butt up against one another. The imbalance appears to put creative forces in motion that attempt to bring more balance to the situation. If there were complete balance in the universe, nothing would happen. Imbalance gets the ball rolling, and probably started it rolling in the first place.

For our purposes, let's say that the hydrogen atom was the first structure. Most structures thereafter are based largely on the energy configuration that we see in that particular atom, that of the male proton and the female electron. Through the actions of the electron, the structure can be increased in size. It is the same with the female and the human family. But as we know, the male initiates the process of human duplication that can only be accomplished through the female. Likewise, in the hydrogen atom, without the proton 'capturing' the electron, there would be no molecule formation or 'duplication'. So therefore, the proton actually initiates the process.

Usually structural design in the human world will include a small but dense nucleus surrounded by 'working' energy close to the nucleus with creative-formative-productive energy (CFPE) embedded within the structure and often on the periphery as well. This CFPE energy will maintain and improve (change) the structure, but some of it may break away and form a new structure or join another structure. You see it in business all the time— some of the original founders often leave and form their own companies.

The universe appears to be built around opposites, what some refer to as the yin and yang. There is matter and antimatter, positive and negative, light and dark, hot and cold, peace and violence, benevolence and greed, production and accretion, and many others. Electrons and protons seem to be in that group also, not only because of their opposite charges but also in the way the energy of each behaves. I think the opposites that many philosophers have written about over the ages regarding the universe and reality are actually the proton and electron energies that lie at the foundation of the atom. The proton appears to be the agent for dispersal and the electron the agent for anti-dispersal.

The electron does not change its characteristics; it is always the same. As the change agent, it is predictable. The change comes about through its connections with other electrons. So it is with the human valence electrons also, the IMPACTS. As we will see, the IMPACTS have maintained their same basic characteristics ever since the San tribe and shaman emerged over 100,000 years ago. Just like the electron, the IMPACTS are mostly predictable and are the initiators of change, though our living world is not nearly as precise and efficient as is the physical world of the atom.

The Greek philosopher Plato thought there were really two realities: the everyday kind that we see all around us and the ideal state. He believed that everyday reality was a poor imitation of the ideal—that the real stuff existed 'out there' somewhere. He also thought that mathematics was part of the poor imitation and thus had no credibility as far as its ability to make sense of the universe. You can see why the Christians loved Plato even though they came 300-400 years after him. His thinking fit right in with theirs, to a degree, but his had no religious connotation.

I believe Plato was just another in a very long line of IMPACTS that started way back with the San, as we will see in the next chapter. Therefore he was always looking for ways to improve his world, and as IMPACTS often do, he first tried to figure out how it all worked. As we will see, the San and their shamans also believed that the answers were to be found in another reality. The San-shaman, or shamans even farther back in time, actually started the idea of a spirit world, which developed into 'heaven' for some.

In the atomic world, the 'between worlds' valence electron appears to stabilize the forces. In the molecular world, carbon provides an excellent balancing agent of sorts because it too is 'between worlds' because of its stable-reactive condition—4 electrons in the outer energy level and 4 more needed for 'satisfaction'. So far, so good. But in the everyday world of human beings, that balance is still very elusive as the 'between worlds' IMPACTS are, yes, bringing everything together just as valence electrons and carbon atoms do, but they are not providing the balance that is needed. If the creative-formative-productive energy isn't strongly embedded at the outset, the structural energy can rapidly take control, and imbalance will reign. The IMPACTS energy is embedded but currently it is not strong enough to provide a counterbalance to the SN, parts of which are behaving recklessly and diabolically. Remember how the maternal instinct is nurturing, innovative, and protective? So too is IMPACTS energy, but its protective mechanisms are currently no match for the SN.

However, nature provides us with a possible solution. As noted the valence electrons provide balance through bonding. Human beings have to do the same; we have to come up with powerful, innovative connecting strategies that can keep the SN in check. Nature has already proved it can be done to a very significant degree. First we have to become aware of the dynamics—then we have to initiate thorough and ongoing bonding actions.

One might say, "Isn't that the purpose of the United Nations and other such organizations?" The UN is controlled by the SN. We need grassroots organizations run by IMPACTS around the world.

In our Milky Way galaxy, massive stars form in proximity to the black hole, displaying a tenacity which seems to surprise many astrophysicists. In our human world, massive IMPACTS-type structures are not being formed or surviving close to the SN. To me, that is a sign that the current state of development of human civilization is askew when compared to the operations of the rest of the universe. We need to find ways to form 'massive IMPACTS stars' right on the doorstep of the SN.

Earth's Development

The universe as we know it is believed to have been born about 13.7 billion years ago with the Big Bang, though it was more of an expansion than a bang. It would be another 9 billion years or so before our solar system would form—about 4.5 billion years ago. Our solar system is located about two-thirds of the way from the black hole at the center of the Milky Way to the edge of the galaxy, situated between two spiral arms (our galaxy is believed to have four) on a 'dead-end street' called the Orion spur. That position may not seem significant, but I believe that life may have started as another antidote to the 2nd Law of dispersal. For that to happen, electrons need room to work, away from the 'unfriendly' environs closer to the black hole.

You may recall from the Introduction that I found that my customers, who are all IMPACTS, preferred to live near a buffer provided by a cul-de-sac, dead-end, body of water, or other natural or manmade barrier. That is the way it is with creative-formative-productive energy—it needs plenty of space away from the SN and its energy.

It was not long after the earth formed that water made its appearance. Five hundred million years after its birth, 90% of the earth was covered by water with tiny volcanic islands dotting the landscape. But the atmosphere was toxic with a carbon haze, and the earth's surface temperature was about 200 degrees Fahrenheit. Many believe that a large percentage of the water arrived extra-terrestrially from comets and asteroids, which is another example of creative-formative-productive energy coming from the periphery.

The peripheral principle may have been at work as well in the beginnings of life on earth. NASA's Stardust mission returned to Earth with thousands of tiny particles from the tail of a comet named Wild 2. These particles contained organic molecules essential to amino acid formation, the building blocks of proteins, which are in turn the building blocks of DNA. The thinking is that such molecules could not have formed in the scorching environment of early earth and would theoretically have had to find another way to get here.

It appears that these building blocks of amino acids are formed in the same dust clouds as stars and planets. Is it all part of creative-formative-productive energy? More than 70 varieties of amino acids have been found in meteorites so it is quite possible that life-forming material arrived here with comets from the distant fringes of the solar system, or with asteroids. Recently a sugar molecule was discovered near the center of the Milky Way galaxy in a gas and dust cloud where new stars were forming. The more we see the peripheral element at work, the less unrealistic it seems that maybe the origins of life were immigrants from afar. Remember, we and the solar system we inhabit are a tiny interconnecting part of the whole. We are isolated from the rest of the universe only in our own minds.

I mentioned the asteroid belt. The reason there are space rocks between Mars and Jupiter instead of a planet is because Jupiter is so massive that it prevents a planet's formation. And even the rocks that remain are a tiny percentage of what existed early on as Jupiter's gravity has flung most of them out of their orbits. That is the way it works with the powerful SN in the human world too. It keeps cohesion to a minimum and tries to allow no space to those who might promote it, unless it serves the SN. Mostly the 'cohesionists', or violators of dispersal, get flung out of the system the same way that asteroids have been projected to peripheral locations by the gravitational forces of Jupiter. I used to wonder why die-hard capitalists hated socialists so emphatically. Now I know. It is dispersal against non-dispersal, the basic battle of the universe.

The Jupiter phenomenon occurs in the international realm also. The big SNs want to become so powerful that their 'gravity' disassembles any efforts to compete with them within what they consider to be their spheres of influence. You will note that we have two sets of vocabularies, one for the physical world and one for the human world. But it should be increasingly obvious that we could drop much of the human vocabulary and only use the physical wording to explain both worlds. The processes and forces are identical.

Seven hundred million years ago, as volcanoes in current-day Panama produced a new land mass which blocked the circulation of warm-water currents to northern latitudes, it is believed that the earth became a snowball of one-mile-thick ice blanketing its entire surface. Bacteria and other very small life forms were trapped under the ice. But gradually over many millions of years, volcanoes caused the breakup of the ice around the world. Volcanoes have played an integral role in the development of the earth and life itself, including human beings. We will see how humans followed the volcano trail as they moved around the world.

About 550 million years ago, the Cambrian Explosion occurred when life literally exploded across the earth, developing the forms more similar to what we see today. Oxygen reached today's levels in the atmosphere, and an ozone layer formed. From 230 million to 65 million years ago, dinosaurs ruled until a six-mile-wide asteroid slammed into Earth, ending their reign and clearing the way for the development of mammals. Another significant peripheral event had changed the structure dramatically. But as you can see, it is unlikely that change would have come from 'inside' the prevailing structure because the dinosaurs had already dominated the earth for 165 million years. It looks like the probabilities finally caught up with them.

If there is a one in a million chance of a huge meteor colliding with earth, and enough time passes, eventually it has a pretty good chance of happening. The structure (the dinosaur world) had become so big and so strong that only something extremely powerful from the periphery could break through the barriers and cause the 'nucleus' to disassemble. The same is true today among humans—'dinosaur' behavior is not easily changed.

The peripheral element, such as the valence electron in the hydrogen atom or the outer valence electron energy level of the carbon atom, is often the stabilizer and the change agent simultaneously. Our moon is another example. The moon is believed to have formed about 50 million years after the earth when the earth was struck by another planet-sized body, sending debris outward which coalesced into the moon. At first the moon was only about 15,000 miles from Earth, compared to 239,000 miles today. When it would rise above the horizon, it would cover about half the sky. But it has been moving away ever since, currently about 1.5 inches per year.

This is a good example of how a peripheral creative-formative-productive agent enters the scene, creates conditions for a new, more stable structural environment, is captured and embedded, and then slowly moves away to the periphery again.

Without the moon, the earth would wobble and not be able to sustain the conditions for life as presently constructed, if at all. We will see the same change effect, and stabilization, from the periphery in human society with the IMPACTS.

Punctuated Equilibrium

We have continued to see the power of the periphery. Let's look at how it influences species development.

The late author and biologist Stephen Gould introduced the concept of punctuated equilibrium as it relates to evolutionary pressures. He said that yes, changes are sometimes gradual, but often major morphology changes come in spurts. Sometimes these alterations can produce new species. This new species might continue basically unchanged for the duration of its existence. An example is the tuatara, a reptile that has seen very little morphological change in 200 million years. Many species reveal the same pattern: develop for a period of time, then remain in stasis for the duration or until another spurt occurs.

Gould claims that the mutations of individual members of the species generally have a minor impact on morphology because a certain homogeneity is reached in the population, and the homogeneity and the environment generally negate the mutations before they can become fixated in the genome. He states that new species and major morphology changes form at the periphery of existing species.

"Selective pressures are usually intense because the peripheries mark the edge of ecological tolerance for ancestral forms. Favorable variations spread quickly. Small peripheral isolates are a laboratory of evolutionary change."[1]

The periphery appears to be a major factor throughout the universe. One such place is on the fringes of the solar system. All of the creative-formative-productive energy (CFPE) did not get sucked in by the gravity of the sun. Trillions of comets reside in the spherical Oort Cloud that surrounds the solar system. These comets are leftover CFPE from the early days of the solar system. The nucleus, the sun in this case, can capture most of the matter and energy, but it has more control over the matter and energy that is closest to it, as is the case in the atom with the nucleus and the electrons. Comets can still affect events, as occasionally they will get knocked out of orbit by a passing star or other massive object or energy and enter the inner solar system. Many of these have certainly hit Earth, and others will in the future. Change resides on the periphery.

Creative-Formative-Productive Energy (CFPE) leads to a Structure and Structural-Nucleus (SN) which in turn captures the Creative-Formative-Productive Energy.

This is the most important point in the book: creative-formative-productive energy (IMPACTS energy) creates or enables the development of a structure, which forms a controlling nucleus. This nucleus then captures (accretes) the creative-formative-productive energy (CFPE) as the primary source of energy for ongoing operation of the structure, including innovation, maintenance, and improvement. The CFPE (IMPACTS energy) becomes the embedded change agent for the structure.

Again, think of a mother and her role in the developing structure of the family. She is creative, formative, and productive—she is having children and building a nucleated structure called the nuclear family. Care of the children requires innovation. Plus, she is working to maintain and improve the structure, and she can change it significantly with another birth. In addition, she is extremely protective of what she has created—her children and family. This is the same general model for a successful entrepreneur and just about any kind of organization.

After the formation of the solar system, it is believed that life started emerging early, 3.5 to 3.9 billion years ago. Early life was simple, mostly bacteria and archaea, both of which are prokaryotes, one-celled organisms which have no clearly defined nucleus. Archaea are often characterized as extremophiles because they can survive in extreme environments such as volcanic hot springs. But they also live in our guts too, as do bacteria.

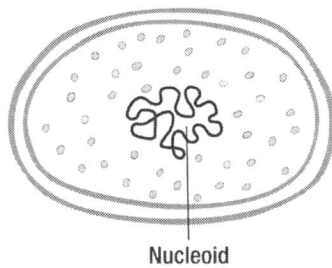

Nucleoid

Prokaryotic Cell – No Clearly Defined Nucleus

Instead of a nucleus, prokaryotes have a nucleoid which is an irregularly-shaped area of the cell where the genetic material (DNA) is kept, itself in basically a circular configuration. Bacteria thrive in water but they are also found in sulfuric acid, oil beneath the earth, and other seemingly toxic environments. This is also instructive in regard to the foundation-laying IMPACTS. The ancestors of modern humans lived and survived all over the globe in every imaginable environment—caves, deserts, jungles, islands, rocky cliffs, 12,000-foot mountains, arctic tundra, and underground dwellings that they carved out of volcanic rock. You name it, and modern humans have lived there at one time or other.

About 3.5 billion years ago, cyanobacteria started a process that formed what could be called 'living' rocks known as stromatolites. These rocks formed in shallow waters all over the globe. The process went like this: Cyanobacteria utilized the sun's energy, along with hydrogen that they extracted from water, and produced their own food, the process known as photosynthesis.

The slimy texture of the bacteria captured sediment, which combined with calcium carbonate in the water to produce layer upon layer of rock. The 'waste' product of photosynthesis, oxygen, was released into the oceans and the atmosphere. Two billion years of work turned Earth into the beautiful blue planet that it is today. But it did more—oxygen paved the way for more complex forms of life because it is extremely efficient in the production of energy. Why? Because of its rapacious appetite for 2 more electrons which are needed to fill its outer energy level. Oxygen combines with just about every element except the noble gases, which are not reactive because they have the required number of electrons in the outer shell.

Stephan Harding explains in *Animate Earth*: "Oxygen is the passionate Italian of the chemical world—its urge to gather electrons is so powerful that it can literally burn up the complex molecules of life, releasing copious quantities of solar energy originally locked up by photosynthesis. Respiration, without which multi-cellular life, such as us, would be impossible, uses oxygen to burn up food molecules in a gradual, controlled way and stores the energy in special molecules such as phosphorus-rich ATP."[2] Nothing happens without the electrons attempting to fill that outer energy level.

An important aspect of the energy-producing process of photosynthesis should be noted. This is also the same process used to produce energy in mitochondria and almost all living organisms. It is the electron transport chain. The process begins in photosynthesis when sunlight heats an electron, and in mitochondria when respiration occurs in the cell. The electron transport chain produces an excess of protons on one side of the membrane. This imbalance of protons drives the production of adenosine triphosphate or ATP, the currency that enables the cell to function. So we can see clearly that electrons get the ball rolling.

Stromatolites are a good example of a structure which would start dissolving immediately if the creative-formative-productive energy were removed. The most important contribution of cyanobacteria was creating oxygen that set the stage for the future. When life started evolving into bigger and more diversified life forms, some of these fed on the bacteria that had enabled their creation. Others utilized bacteria to aid in their own existence, as we do. While a portion of the creative-formative-productive life forms (bacteria) got captured, others remained free, working for the greater good as they had before. We will see the exact same thing happening among human beings. Some of the IMPACTS and their energy get captured by the SN and its supporters, while other IMPACTS stay on the periphery where they avoid capture and continue their work for the greater good. We have already seen examples with Linus Torvalds, Tim Berners-Lee, and others.

About 1.5 to 2 billion years ago, life took a gigantic step. A cell formed with a nucleus though the exact way it happened is unclear. Nucleated cells are called eukaryotes. A virus could have taken over a cell, becoming the nucleus, and then capturing bacteria to be its energy producer. Or an archaeon could have combined with a form of bacteria in a process known as endosymbiosis. Or it could have happened another way. No matter—the important thing is that a nucleus was formed, capturing bacteria as its source of energy. These bacteria morphed into what we call mitochondria, which are found in most living cells. Not only do mitochondria produce energy, they also regulate cellular metabolism and perform a host of other functions. IMPACTS are the mitochondria of the human race.

Eukaryotic Cell

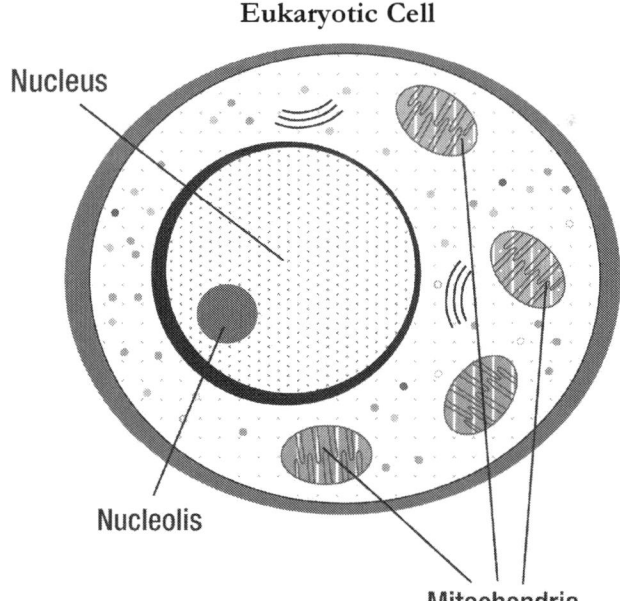

The eukaryotic cell has many different parts called organelles. Each organelle has a different job within the cell.

The same utilization of an existing energy source occurred with plants. The chloroplasts that enable photosynthesis within plants are captured cyanobacteria, which started the whole oxygenating process.

When eukaryotic cells formed, the nucleus and mitochondria basically kept their own DNA. Mitochondria had to produce energy for the cell, and the cell was dependent on a peripheral agent to supply its energy needs. They were locked together, much more than are the nucleus and electron of an atom. An electron can 'escape' if it receives the right amount of energy. Mitochondria are not going anywhere. Life forces replicate the model of the atom, but life forms 'structuralize' the forces, making change difficult.

There are far fewer genes in the mitochondrial genome than are seen in its bacterial cousins because some genes have been lost and others have been absorbed by the DNA of the nucleus of the cell. That absorption increases the control of the nucleus over the mitochondria and entwines their activities more tightly. They are joined as one.

The exact same thing has occurred in the human world as the SN has captured a large percentage of the innovative, energy-producing IMPACTS. It too is like they have different DNA. But just like the nucleus of the cell, the SN has been steadily accreting the 'genome' of the IMPACTS; hence, the SN-IMPACTS. The P-IMPACTS largely retain the original IMPACTS 'genome'. As mentioned, the subconscious goal of the SN is to capture all useful IMPACTS energy within its sphere with the remainder essentially being ignored or discarded—or destroyed if it is a threat.

Cells might contain a single mitochondrion, or they might contain thousands of mitochondria; it depends on the needs of the cell. If more energy is required, mitochondria will grow and divide by binary fission, similar to bacteria. When less is required, they become inactive or simply dissolve. Mitochondria can also fuse with other mitochondria when more energy is required. Also, new mitochondria will sometimes be synthesized in areas of the cell rich in certain proteins and polyribosomes.

It almost seems as if mitochondria have their own little brain. But the more we look at life, the more of that we see—many parts appear to have their own brain. I think it starts with the electron and is then manifested in the valence electron as it attempts to rectify its imbalance. It is trying to solve a problem. Early life forms were also trying to solve problems as we saw with cyanobacteria. Perhaps electron energy in general across the universe is problem-solving, or making-well, or balancing energy.

When atoms formed about 380,000 years after the Big Bang and electrons were seized by protons, the universe essentially went dark. Ninety percent or so of the atoms were hydrogen and still are today. The remaining atoms were mostly helium. Before atoms were formed, free electrons scattered electromagnetic radiation in a process known as Thomson scattering. When electrons became tied to protons, electrons and radiation uncoupled with the radiation becoming the cosmic microwave background.

Galaxies and stars started forming during the Dark Ages, which lasted several hundred million years. It was probably radiation from stars that re-ionized the universe, knocking electrons away from protons and forming plasma once more. Some however believe that it could have been radiation from black holes that did the job since black hole radiation is so much more powerful than that produced by stars. Maybe it was a combination of both.

During this time, we get a glimpse of what could possibly be the balancing role of electrons in the universe. Astrophysicists believe that there were three distinct temperatures during the Dark Ages: spin temperature related to the spin states of atoms, the kinetic temperature caused by the motions of atoms, and the radiation temperature of the microwave background photons.

Abraham Loeb, in <u>Scientific American Magazine</u>, October 16, 2006, states: "As space expanded, both the gas and the radiation cooled . . . the gas would have cooled faster, but initially a small residual number of free electrons left over from the formation of hydrogen atoms counteracted this tendency. These electrons acted as **middlemen** [emphasis added] to convey energy from the microwave background to the atoms, keeping all three temperatures equal. Ten million years after the Big Bang, however, the electrons faltered in their role because the microwave background had become too diluted. The equilibrium between gas and radiation broke down, and the gas started to cool rapidly."

The Dark Ages of the universe were thus initiated by electron capture, and that is also when the structures known as galaxies started forming. There are two lessons here for the human world. The first is that structures grow when the electron IMPACTS are captured. The second is that dark ages also occur when the electron IMPACTS are captured. The danger for the future of human civilization is not the physical subjugation of the IMPACTS by the SN—it is mental subjugation, or the victory of homogeneity over anti-status quo thinking. That would be the victory of the 2nd Law, human-style. As is so often the case in the human universe, what appears to be victory is actually defeat.

Bacterial remnants, chloroplasts and mitochondria, are still performing the two critical functions that are so necessary to life on earth today: photosynthesis and energy-production for the cell. It is part of the great recycling, frugality, and efficiency of life in its utilization of energy. Life does not reinvent the wheel—it uses what is already on the shelf, efficiently conserving energy. It also demonstrates the incredible duplicating power of even small amounts of creative-formative-productive energy, or IMPACTS energy. It starts with the tiny valence electron which has tremendous ability to shape events—through bonding.

Bacteria help us digest our food and fend off bad bacteria. We could not survive a day without them. Amazingly, there are more bacterial cells on us and within us than there are human cells. Is it possible for this inter-species dynamic to be operating within the human species—two very different elements working together, almost like two different human species? Yes!

If you have ever watched water filling up an aquarium, searching out and covering every little crack and crevice, wrapping itself around all the rocks and stones, that is a good picture of the attitude of the IMPACTS, making sure nothing is missed and everything is covered. That also appears to be the attitude of life itself, finding a way to survive and thrive, to optimize potential, in almost any environment, even extreme ones like bacteria living in sulfuric acid or in oil beneath the earth's surface. Life may actually have started in such conditions and evolved as environmental conditions changed.

Left Brain-Right Brain

The human brain appears to be a good example of the creative-formative-productive energy and structural-nucleus (SN) duality at work. It is the atom manifested along with the 2nd Law and its antithesis.

Let's draw a line to represent human behavior. On the left end we will put engaged with the external environment, and on the right we will put autistic, or not engaged with the external environment. In our society, the engaged behavior is considered normal and valued, and the unengaged abnormal because it is not seen as contributing value, and today that generally means economic value. Behavior and/or thought patterns that do not conform to the prevailing accepted model of modern mental health are likely to be considered abnormal, and therefore 'treatment' may be prescribed. The goal is to get the person to function in a way that fits the definition of the healthy model, which includes contributing to society in accepted manners.

Engaged with the external environment	Autistic or not engaged
Normal and valued	Abnormal – not valued

But again, that is part of the continuing trend that developed after the advent of agriculture when the emerging SN 'naturally-selected' the kind of people that it wanted. It saw little value in those who could not aid in the production of valuable goods, needed services, structures, and works of art that enhanced the throne, or could not fight. Much of what we are as modern humans, however, appears to have originated from the autistic area of the spectrum. And that also appears to be the origin of the IMPACTS profile. But that should not surprise us. We are continuing to see that creative-formative-productive energy emanates from the 'loosely engaged', as evidenced in the hydrogen atom with the lone electron. We will learn more about our origins in the next two chapters when we talk about the San and San-shaman.

Leonardo da Vinci, who possessed one of the greatest minds in recorded history and perhaps one of the greatest ever, would certainly today have been classified as abnormal and in need of treatment. Among other psychiatric difficulties, he may have had Asperger syndrome, a high-functioning form of autism that has 'afflicted' many of the truly great contributors to human civilization, including possibly Einstein, Thomas Jefferson, Archimedes, and millions of others. Da Vinci cared little for traditional education and had a habit of not finishing what he started, which is not uncommon for people like him who want to accomplish a million different things. Were he living today his genius energy would most likely go to waste. How many da Vincis are being cast aside today in the SN quest for homogeneity? Homogeneity has no space available for loose engagement. Those who are loosely engaged are not contributing enough according to the SN, and therefore, they end up on the periphery as castoffs.

Let's look at the two hemispheres of the human brain for a moment to see how the makeup of the brain fits in with our SN versus IMPACTS theory. These are some traits generally associated with each hemisphere.

Left-brained
- Analytical
- Objective
- Detail-oriented
- Present and past
- Math and science
- Responds to verbal instructions
- Processes information from parts to whole—takes pieces and lines them up to form a sort of logic—sequential—rational
- Looks at differences
- Planned and structured
- Controls feelings
- Prefers ranked authority structures
- Draws on previously established, certain, and organized information
- Realistic and rigid—reality-based—facts rule
- Forms strategies
- Practical
- Safe

Right-brained

- Responds to demonstrated instructions
- Imagination rules
- Often solves problems with intuition and hunches
- Looks for patterns, similarities, and configurations
- Fluid and spontaneous
- Prefers elusive, uncertain information
- Prefers open-ended questions
- Freer with feelings
- Prefers collegial authority structures
- Connectedness is important
- More flexible
- Processes information from whole to parts—sees the whole picture first
- Appreciates art-music-culture
- Philosophy and/or religion may be important
- Present and future
- Non-verbal
- Uses metaphors and analogies, symbols and images
- Emotional
- Random yet holistic which suggests experimentation in an attempt to connect dots
- Synthesizing
- Subjective
- Empathic
- Drive-oriented
- Creative
- Global—Sees the BIG Picture
- Can "get it" (the important thrust of the message)
- Appreciates
- Believes
- Looks at possibilities and potential
- Impetuous
- Risk-taking

The right-brain clearly looks more female, creative, and loosely-engaged while the left-brain appears more male, structured, and engaged. One could deduce that the left-brain's interest is in power and control. The descriptions even suggest military aspects. As is the case in the atom, the two forces are separate, but in the brain the two hemispheres are connected through a large body of nerve fibers known as the corpus callosum, which is generally thicker in females.

Looking at the list above, I can see IMPACTS traits on both sides though they are certainly more prevalent on the innovative right-brain side. The IMPACTS bring things 'into being', usually as they work with real 'stuff' in a hands-on manner, and they deliver results. Therefore, there has to be a solid structural component to the profile. IMPACTS women usually have a strong male side, and IMPACTS men have the structural side and a strong female side. Again, the model of the atom is relevant where it is female energy, the valence electron, which 'remakes' the structure and delivers results.

Today's world is a male structure with clear instances of nucleus- or SN-formation. Pre-agriculture was more prokaryotic in makeup with only a loosely-defined nucleus—more right-brained. In our current world, the right-brain is embedded in service to the left which we can see with the capture of the IMPACTS by the SN. The IMPACTS are utilized to enhance the structure—to make it stronger, more efficient, more productive, and more appealing. It is true for any structure—a business, a country, the theatre, the Super Bowl.

Autism

There has been a surge of reported autism cases in the past few years. Simon Baron-Cohen, director of the autism research center at Cambridge University in England and author of *The Essential Difference: The Truth About the Male and Female Brain*, has some theories. But first a little background.

The corpus callosum, being thicker in women, may be the reason that, according to a study from Yale, women are more likely to activate both hemispheres of the brain when performing language tasks whereas men generally activate only the left hemisphere. Men seem to do better in math and mechanics overall, and women score higher on people issues such as emotion recognition and social sensitivity, and higher also on language ability. But IMPACTS men and women will often be the reverse from the model above—IMPACTS women are often very adept in math and mechanics, and IMPACTS men are frequently more aware of nuances in the social environment than are other males.

Baron-Cohen believes that on average females have a stronger drive to empathize and men a stronger drive to systemize. Identifying how a system works may allow you to control it or predict its behavior. Empathy allows you to have a better understanding of the needs of others and to respond appropriately. Again, IMPACTS men and women often seem to have both—empathy and a systemizing-orientation. IMPACTS are closer to androgyny than the rest of the population; therefore, they will have discernible male and female elements.

It has also been found that the amount of prenatal testosterone produced by the fetus is a determining factor in the kind of brain that is produced. If a male's testosterone levels are low he may have more of a female brain, and if a female's testosterone production is high she may have more of a male brain.

Baron-Cohen believes that the autistic person has a particularly strong drive to systemize and little drive to empathize; in other words, an extreme version of the male profile. His research has found that both parents of autistic children tend to be strong systemizers, and that the father of each parent is more likely to be a systemizer. So systemizing appears to be in the genes.

In the vast majority of cases, autism is found in males. Why would that be the case? Perhaps the 'autism gene' has value for humanity in that it houses immense innovation as demonstrated throughout recorded history, particularly from those with Asperger syndrome. Nature will keep that which enables survival. That would also explain why males come up with most inventions—because of the autistic remnants. Don't get me wrong. It is clear that I am a strong supporter of females, but I am trying to look at all of it with as much science and as little bias as possible.

Instead of extreme maleness there may be something else at work in autistic males—a stripped down male ego. That may be what we are seeing—the absence of the left-brain, the control center. That would also account for the frequency of right-brain innovation.

Generally there appears to be little 'macho' behavior in autistic males, another sign of an absence of the male ego. So-called low-functioning autistics appear to literally be stuck in 'no-man's-land'—in another world. But as we will see in the next chapter, 'another world' is what modern humans were built upon. Nature has probably kept autism through natural selection because the autistic spectrum houses much of the innovative energy for our species. Therefore, I believe the IMPACTS profile emanates from the autistic spectrum though very few IMPACTS are autistic. The small but significant number of people with autism may be paying the price for the existence of modern humans. More about this in the next two chapters.

Business

A business is a good example of the creative-formative-productive energy (CFPE) leading to the formation of a structure and structural-nucleus (SN). First there is a creative-formative-productive agent or agents who start the business. In the beginning, the business may be small with seemingly boundless energy and no real signs of a nucleus. But gradually the business grows (hopefully), and a structure forms on top of an infrastructure along with a nucleus which handles day-to-day operations. Usually, this original CFPE, the entrepreneur, stays with the business for a considerable time, making sure the energy is embedded within the structure.

To continue optimizing the realization of the potential of the business, there must be continual assessment, and continual creation and delivery of needed elements. This means being open to new ideas, continually learning, encouraging innovation, exercising restraint when required and boldness where appropriate, letting passion take over in measured ways, and encouraging freedom of thought and expression in the entire working environment.

Let's look at Steven Jobs, an early adversary of Bill Gates, who like Gates while he was still with Microsoft, is a large part of the creative-formative-productive energy and the structural-nucleus

(SN) of his company, Apple. Did you know that Steven Jobs was adopted? His adoptive mother had a high school education and his adoptive father did not have that. His biological father was Syrian and a professor and his biological mother was American Caucasian and a speech pathologist. Many IMPACTS are adopted. Why would that be the case? Because IMPACTS parents are usually trying to do the right thing, and though giving up a child for adoption is never easy, an argument could be made that sometimes it is the right thing to do.

As noted earlier, IMPACTS are often found around serious challenges or rifts, sometimes semi-detached or detached from the normal structure, and frequently 'between things'. These are manifestations of the peripheral element. The valence electron is trying to fix the imbalance that exists; IMPACTS are trying to do the same. Often, important bonding has been severed in the lives of IMPACTS, and they are trying to create a new world with new bonds. As we have seen, they will also place themselves in positions where others are experiencing serious bonding issues, such as emergency care, disaster relief, special-needs situations, or adoption. The base of the IMPACTS profile rests upon bonds and bonding—how to repair them, how to strengthen them, how to create them, how to go forward when they have been severed, and very importantly, how to help others do the same.

After completing high school in the Silicon Valley of California, Jobs attended Reed College, a small, highly respected liberal arts college in Oregon. He dropped out after one semester for lack of funds, but continued auditing various classes. A couple of years later he backpacked around India in search of 'spiritual enlightenment', returning to California as a Buddhist.

While working at Hewlett-Packard during a summer in high school, Jobs met Steve Wozniak, who would later become his business partner. You will recall that HP was a company that started with copious amounts of IMPACTS creative-formative-productive energy and attitude, but little else. IMPACTS will often work at an IMPACTS-type company before they go off and start their own business. You will find IMPACTS in clusters, attracted to the same types of companies, organizations, and values—even living environments.

Why would computers attract the eyes and brains of IMPACTS? In a nutshell, they are efficient. They safeguard knowledge, they enable it to be duplicated and proliferated endlessly, they increase productivity, they aid bonding among people, and they help solve problems. IMPACTS can do more of what they want to do because of computers. Computers are in the same family as robots. That is why when you see an emphasis on robotics, you will find IMPACTS. Japan is a good example as it is the world's leader in robotics and is an IMPACTS-laden society.

Jobs and Wozniak started Apple, makers of Macintosh computers, but after a few years, in 1985, Jobs was fired from his own company. The embedded creative-formative-productive energy was gone and the company suffered, as is often the case when the founder leaves for whatever reason. In 1997 he returned when Apple bought one of his companies. Apple has recovered, largely due to the efforts of Jobs.

We saw that Gates's parents contributed their time and energy to the community, and that Jobs's biological mother was a speech pathologist. Nurturing genes and innovative genes generally reside together though there will be different levels of manifestation within a family. One child may want to save the world, another may want to be a successful model, another may want to be an entrepreneur, and some won't care one way or the other. You never know. A lot depends on life's twists and turns. But generally if you find genuine caring and compassion, you find innovation close by. Think of it this way—if you really want to help people or animals or anything else in need, you will find ways to do it, even if you have to invent them. And that is often exactly what happens with IMPACTS.

You can see the swirling activity around Steven Jobs as he attracts followers and creates mostly positive upheavals. Electrons have a characteristic called spin which causes them to act like tiny magnets. IMPACTS are the same. Their 'always in motion' energy and mental makeup often create movements around them, sometimes big and sometimes small. Obvious examples are Martin Luther King, Jr. and Mother Teresa, but the same applies to people in any field of endeavor whether business, art, medicine, science, non-profits, sports, religion, or anything else.

Note the 'regular' people that were around Steven Jobs early in life, the intercultural aspect, the nurturing-caring genealogy, the early challenging environment (bond severance), the reaching out from others to form new bonds, the 'off the beaten path' issues, technological innovation, artist-orientation (Macs are loved by the artistic), recovery from bitter disappointment and further bond severance, and the unceasing creative-formative-production and delivery. Some IMPACTS just keep re-creating themselves because that is what they have to do.

Everyone of course does not love Steven Jobs, or Bill Gates for that matter. IMPACTS are not in a popularity contest. They are here as they see it to get things done and hopefully change the world for the better. Worrying about whether people like you or not can detract from that mission.

Leptons such as electrons are not affected by the strong nuclear force. It is the same with people; the IMPACTS have no natural connection to the SN. They are similar to visitors to the party; the SN has to convince them to stay but the IMPACTS are not sure if the party is right for them. The more the SN can entangle them, the more apt they are to stay. These machinations are rarely conscious; it is just the way a hierarchal structure operates as it tries to capture its sources of energy. We can be sure that the same process occurred as the nucleus of the cell tried innumerable strategies to keep the energy-producing bacteria (eventually mitochondria) inside the cell. It no doubt took quite a while but it obviously succeeded.

Black Holes

A structure develops its own momentum. Look at our galaxy, the Milky Way. It becomes more powerful with each star that is created. Without stars, there would not be much of a galaxy. The stars give the universe its light, and they give it its diversity too through the production of new elements, most of which are released through supernovae. Stars, though they make up most of the mass of a galaxy (except for dark matter), appear to have absolutely no control over it. They are stuck though occasionally one does escape. The super-massive black hole at the center 'manages' the galaxy—everything revolves around it. It is very similar to what we see with the United States. The 300 million people or so are managed by a tiny few. Does the general population have any control over the structure? Not much, though the SN would have us believe that we have significant control.

We have noted how the SN appears after IMPACTS energy (CFPE) has started laying a strong foundation. The black hole at the center of our galaxy may have formed after the demise of a massive star, as millions of other black holes have also formed. So the black hole is the opposite of a star—it is its death—but it depends on stars for its survival. It has lost its ability to create and produce energy, and therefore it must accrete energy in order to maintain its place in the universe. While it grows in strength from eating stars, it also grows in strength as they are produced, enlarging the galaxy. This enables it to potentially capture other galaxies, which it will also feed upon, and thus become even more powerful, all without producing a thing. It is using energy from 'others' to enlarge its domain.

A recent study found distant galaxies, some 11 billion light-years away, 'going through puberty'—hot and chaotic but about to stabilize. Lead author James Geach of Durham University in England said that the reason for the chaos "is due to the violent processes occurring in the galaxies, black hole growth, starbursts, mergers. They're having a final 'tantrum' before they're done growing and then 'passively' evolve to the present day . . . These could be the signal of galaxies coming of age." These galaxies are far larger than our own Milky Way. Hence, their formation may have been different in some ways.

The black hole in the 'pubescent' galaxy is growing with the developing galaxy. As things calm down, the black hole will assume its lofty position. The same process occurs on Earth with human organizations. The IMPACTS lay the foundation, an SN grows and develops with the CFPE, and then the SN takes control of the structure and the CFPE, using it to solidify and often expand its position.

Such is the universe, two distinct forces at work, consumption and production—further manifestation of the energies of protons and electrons, and of the 2nd Law and its violators. We see it in the people world too. A segment of the population is the main consumption center, sucking up the energy and production from the producers.

Another aspect of black holes reveals more of their paradoxical role in the universe. As incoming matter interacts with the magnetic field of the black hole, tremendous jets of energy can be released into the surrounding cosmos. These jets warm the environment and dramatically limit star formation. You will recall that stars form in cold molecular clouds of mostly hydrogen gas. The black hole is the 'cosmos planning department'. It is extreme male energy and it wants to 'control the neighborhood'.

The same thing happens on Earth. The SN is deeply concerned about what is essentially IMPACTS star-formation outside of its current sphere of influence. Therefore it attempts to enlist IMPACTS and others in its efforts to control its neighborhood, which really means reducing the potential of other SNs and their IMPACTS. As we have seen, that is the true intent of the 2nd Law of Thermodynamics—to eliminate potential, which means to eliminate different concentrations of energy. Many domestic IMPACTS can see through this gamesmanship and want no part of it. Yes, they want to create and produce but they do not want to strengthen the SN. That is why many P-IMPACTS remain on the periphery, away from the SN-empowering zone.

It appears that early in a galaxy's formation, the black hole at the center consumes the equivalent of ten average-size stars per year. As the galaxy develops, it eats less. The size of a galaxy largely matches the size of its black hole; the bigger the black hole, the bigger the galaxy. The same occurs in human institutions. Early on, the organization will require many IMPACTS 'stars'; e.g., the founding fathers of this country or the creative-formative-productive people around a business startup. But as the structure develops, male structural energy will generally take control and solicit less input from IMPACTS. This is not because less input is needed; it is just the way the two energies interact after the IMPACTS have enabled SN formation. IMPACTS energy is embedded in varying degrees in organizations, but generally it is held at bay. It is needed and feared at the same time.

As the structure develops further—any structure—it becomes more and more difficult for a creative-formative-productive agent to influence the whole, just as the formation of a new star does not put a dent in the power of the galaxy and its black hole. What if Thomas Jefferson came back tomorrow and ran for President—would he be elected? Not a chance. The structure, the U.S. government, has moved in a different direction. Besides, Jefferson espoused a 'revolutionary' attitude towards government, believing it required continuing reinvigoration. No such attitude would be acceptable today. Jefferson would be wiretapped, his email would be read by the National Security Agency, and many other aspects of his personal life would make him 'unworthy' to serve.

Barack Obama has significant IMPACTS energy. Which will be affected the most—he by the structure or the structure by him? He will be far more affected by the structure, which has become well established over the past 200+ years.

Creative-formative-productive energy obviously has a strong influence in the early developmental stages of any structure, but if and when a nucleus forms, dispersal forces are initiated. We saw it in the first eukaryotic cell as the nucleus captured bacteria, now called mitochondria, to be used as the energy producers for the cell. In the process, it broke the natural cohesion of the bacteria, thereby forcing the bacteria to work for the nucleus rather than for their own bacterial colony. It did to bacteria what Jupiter does to the asteroids—it kept them separated. But even inside the cell, if more energy is required, the bacterial mitochondria will merge together or divide and increase in population. They have lost their independence, but they are still doing their job of producing energy.

Bacteria were not brought into the nucleus because their role was to produce energy, not to direct processes. Therefore, they were kept in the cytoplasm. The nucleus did, however, gradually accrete some of the mitochondrial (bacterial) DNA, ensuring control.

In the family, the mother creates the structure and usually becomes the primary part of the nucleus at the same time, and slowly the baby grows and moves away from her and the father. Their influence is strong in the beginning but it gradually begins to wane. However, through strong creative-formative-productive energy, the mother along with the father can prepare the child for life in another structure beyond the family, a structure over which the parents have very little control. That structure today is nucleated, hierarchal, and male-dominated. The nuclear family has prokaryotic aspects and eukaryotic aspects while today's SN world is primarily eukaryotic. People are always looking for that prokaryotic space in their life where boundaries and roles largely evaporate, where they can relax and 'be themselves'.

Just as the mother's influence diminishes as the structure of the family develops, so it has been and continues to be the case with the IMPACTS and human civilization. Their creative-formative-productive energy is powerful, and they are still exerting an indispensable influence on the structure of human civilization. However, like the mother, the ability of the IMPACTS to affect events in certain areas becomes diminished as the structure and its structural-nucleus grow.

Comparison of Valence Electrons to IMPACTS

Let's briefly look again at the similarities between IMPACTS and valence electrons. Keep in mind that IMPACTS and valence electrons are anti-dispersal forces.

- Both environments are basically circular.
- The electron field is huge. The same is true for the IMPACTS; they have few borders or boundaries.
- The valence electron is on the periphery where many IMPACTS also reside.
- The valence electron, if it is being shared with another atom, is responding to a critical need for balance which will bring symmetry. IMPACTS are also working for balance.
- Both are optimizing the situation or attempting to.
- Both bond well with like-energy though they both have a natural independence.
- Both are creative-formative-productive agents of the structure.
- Both can transform their environments.
- Much of what happens in the physical world happens because of valence electrons; the same with the IMPACTS in the human world.
- The strong nuclear force does not attract the electrons; electrons are held to the atom by the opposite charge and most likely by Einstein's general theory of relativity which says that mass and energy bend space-time. It is the same with the IMPACTS. Their ties to the SN can be tenuous; there is no natural affinity there. But the power of the SN sucks many of the IMPACTS in toward the center.

- Society can be thought of as having IMPACTS energy at various distances from the SN, and at various strengths, in much the same way as happens in an atom where there are multiple levels of electron energy.

- There is a strong nuclear force holding the nucleus together, including the individual protons and neutrons. Electrons have no such force holding them together. They are more independent, like the IMPACTS.

- Electrons like to be paired with another electron. Likewise, IMPACTS are 'synergized' by working and socializing with other IMPACTS.

- Valence electrons are mobile; so too the IMPACTS. This mobility allows for continual dissemination of creative-formative-productive energy and ideas. Art and culture are excellent examples.

- Valence electrons are the glue that holds molecules together; IMPACTS are the social glue.

Wolf pack

Over millions of years, in order to survive and not spend endless time fighting over food, wolves have developed the structure of a small pack with clearly defined roles. An alpha male and female lead the pack, which includes their offspring and occasionally other wolves which have joined the pack for one reason or another. A beta male or female, second in the power structure, is the glue for the pack. He or she helps settle conflicts (peacemaker), cares for the new pups when needed (caregiver), and looks out for the needs of the group that might be missed by the others (protector). Betas may or may not try to assume the alpha role in the course of their lives.

The rest of the pack constantly struggles for a position in the pecking order but the lowly one, the omega, has the worst of it by far. The omega cannot eat until the alpha gives it permission. Sometimes permission is not granted even though the alpha may be directly related to the omega. The omega's existence is one of constant abuse, and he or she is expected to be totally submissive to the others, just as the others are expected to submit to the alphas. But roles can change. Nothing is set in stone. An omega can run away, find a mate, and become an alpha.

Sometimes, alliances within the pack will try to overthrow the alpha, and these efforts will succeed on occasion. Wolves will remember how the alpha behaved towards the pack and will sometimes banish him. It does not sound that different from human society, does it? We kid ourselves by thinking we are so different.

The humans with the most IMPACTS energy are going to be similar to the beta wolf. They are going to be the glue that holds human society together and the world society together as much as is possible at the moment. IMPACTS are the innovators, the caregivers, the peacemakers, the searchers for the truth, those who are not satisfied with the status quo. In other words, IMPACTS will try to reorient the energy field when they sense that it is not working optimally, which it never is as far as they are concerned. That is why IMPACTS continually work for improvement.

In the next chapter, we will take a close look at modern humans' ancestors, the San tribe of Africa, and explore how the IMPACTS profile developed. All of this is simple, following very basic patterns and principles. Like Europe before Copernicus, it only appears difficult because the wrong model is being taught and 'thought'. The Church is still in control though now it takes the form of an overpowering SN. SNs are not identifying the dynamics operating in society and around the world correctly because they have no interest in doing so. Many are just trying to take what they want and create the reality they want.

Summary

If we forget what we have been taught and look at ourselves with the same objectivity we would use in studying a pack of wolves, I think we will discover some new patterns that we have not seen before—in ourselves and in the universe at large.

Processes and results of IMPACTS energy and of the work of the IMPACTS can be characterized in different ways: renewal, recovery, renaissance, rebirth, transformation, unwell to well, unusable to usable, the best it can be—all in the quest for balance and optimization, which also means realization of potential. It is almost always about new beginnings and new bonds.

As we go forward, I would like for you to remember the model of simple bacteria, how they created the conditions for more complex life forms and were then often captured by the structure that they had enabled to develop. We will see this again and again throughout our discussion. A creative-formative-productive agent begins the process of structural development. Then a structural-nucleus (SN) forms which captures the creative-formative-productive agent, using this innovative IMPACTS energy to maintain and improve the structure. Keep the mother of any species in mind—she creates, forms, produces, innovates, maintains, improves, changes, and protects the structure, her family.

Everything we see is a structure, from the universe on down to the simplest hydrogen atom. Every structure needs an embedded change agent or agents if it is to deal with the changing needs of the environment, and the embedded change agent for structures appears to be their creative-formative-productive energy (CFPE). This is a key point to understanding what the IMPACTS are all about. IMPACTS behave just like the dissatisfied and unfulfilled valence electron in the outer energy level. They seek, connect, share, expand the structure through their efforts and actions, and try to realize the potential that exists in the surrounding environment. In the process, IMPACTS transform the structure and are transformed themselves, from the catalyst to the glue that holds the new structure together. Again, think of the mother. She is the creative-formative-productive energy, and then she is the glue.

IMPACTS deliver results. In that regard, they are like the photon, the force-carrier particle for the electromagnetic force. And just as the photon has no charge, the IMPACTS can generally see both sides of the coin too. This helps them settle disputes and get along with most people, which enables survival and further DELIVERY of whatever is critically needed. But of course, even the most justice-oriented IMPACTS will sometimes find that their best efforts at peace and conciliation will be defeated by the intransigence of others. And sometimes they will discover that they too will have to fight to protect themselves and their loved ones. That is generally the last and most unattractive option because the sincere desire of the IMPACTS is usually to understand all sides and to build bridges between and among people.

Footnotes

[1] Gould article, 1977
[2] *Animate Earth*, Stephan Harding, p 94.

Chapter 4

The San Tribe

It appears that human life emerged in a physical environment in East Africa very similar to that in the sea where life itself may have evolved—around volcanoes and volcanic vents. This should not surprise us; early modern humans behaved in many ways very much like prokaryotic bacteria. They laid the foundation for a nucleated society just as bacteria did for the eukaryotic world. We will continue to see such manifestations of modern humans' integral relationship with nature if we will only keep our eyes open.

Let's be careful as we try to determine the origins of modern human behavior and what connects people, past and present, but let's not be too careful. Far too often common sense is sacrificed as the focus hones in on minutiae. Let's keep in mind that science is about the discovery of new knowledge, or maybe I should say the awakening to it. Ultimately, it is about ascertaining the truth as much as is possible at that moment. The paradigm-enforcers of society stand ready to define the truth for you. They will also tell you if your discoveries are legitimate or not. I would recommend that you try to figure it all out for yourself, with the help of others who are equally open-minded.

Human beings have been around for 5-7 million years, but of course there were many different species before we emerged. Hominids came first, of which there were several species, including Australopithecus afarensis, the species of the famous Lucy skeleton found in 1974 in Ethiopia. Lucy is believed to have lived 3.2 million years ago. Then about 2.5 million years ago, rudimentary tools started being used and hence, a new genus emerged, that of homo. Some of these species included homo habilis, homo heidelbergensis, homo neanderthalensis, homo ergaster, homo erectus, and others.

Homo ergaster may have lasted a million years, using the exact same tool the entire time, never even putting a handle on it for more control and leverage. No innovation, or certainly very little of it. That is what marks the difference between the other human species and us—innovation. But modern humans have only been here about 200,000 years or so. That means another 800,000 years before we can equal homo ergaster's longevity. Will we make it? As a friend of mine put it, this human species is morally expensive as it relates to life and the earth. We are innovative, yes, but we are destructive too, far more than all of the previous human species combined many times over. In fact, there is really no comparison. Something is askew in this species—imbalance reigns—and this book is partly an attempt to uncover the causes. Hopefully, with new knowledge and new ways of viewing the past and present, we can start the process toward creating more balance.

In his book, *A Short History of Nearly Everything*, Bill Bryson relates that there is an ancient tool manufacturing area in southern Kenya known as Olorgesailie. Geologist J. W. Gregory discovered the site in 1919, but excavation was not begun for over two decades when it was undertaken by the husband and wife team of Louis and Mary Leakey. It appears that the site, which was next to a large lake while in use, was utilized for about a million years up until 200,000 years ago. Axes were made from quartz and obsidian that had to be carried to the site from about six miles away. Organization was evident with some areas devoted to crafting new axes and other areas used for re-sharpening blunt ones.

Who were the people who worked here for a million years? No human bones have been found on the site that would offer clues. It could have been an early aggregation site where various groups assembled, something that occurred frequently as modern humans developed. Aggregations were a way to keep strong human bonds intact and to build new ones.

Two hundred thousand years ago, when it is believed that this group met its demise, is about the time of the birth of homo sapiens. The lake appears to have dried up about the same time that the Great Rift Valley transitioned into the challenging environment it is today. A new species could possibly have formed on the periphery out of this tool-making group as the changing climate required new adaptations.

No matter how early homo sapiens originated, it would take another 50,000 years or so for the transition from homo sapiens to homo sapiens sapiens, or anatomically modern humans, to be completed. It was 'knowing human' to 'extra-knowing human'. Some refer to this transition human as homo sapiens idaltu. Homo sapiens sapiens and homo sapiens idaltu are subspecies of homo sapiens.

Let's look briefly at the geography and topography of the cradle of human beings. This birthplace will tell us a lot about the travels of human beings as they spread out around the world. The Great Rift Valley is about 4,000 miles long and averages a 30-to-40 mile-width, running from Lebanon and Syria in the Near East southward through eastern Africa to Mozambique. It is one of the most extensive rifts on the earth's surface. A rift is a crack in the earth's crust caused by the movement of tectonic plates.

The Great Rift Valley is often lined by stone cliffs hundreds of yards high. From its beginnings in the Beqaa Valley in Lebanon, it runs through Israel, becomes the Jordan River and the Dead Sea, the Gulf of Aqaba, the Red Sea, and then becomes part of Africa at the Afar Triangle or Danakil Depression of Eritrea adjacent to Ethiopia. The Afar Triangle appears to be a triple junction where three tectonic plates are pulling away from each other.

This pulling action has already separated Saudi Arabia from the Horn of Africa, forming the Red Sea and the Gulf of Aden. Part of the rift continues eastward underneath the Gulf of Aden and then as a ridge under the Indian Ocean. Another fork heads in a southwesterly direction as the Great Rift Valley, splitting the Ethiopian highlands. It then forms two sections on either side of Lake Victoria in Kenya, the Western Rift and the Eastern Rift. Lake Victoria, the second largest fresh water lake in the world, is shared by Tanzania, Kenya, and Uganda and forms the headwaters of the Nile River.

The Western Rift, home to some of the world's deepest lakes, is also lined by some of the highest mountains in Africa. Evaporation, shallow waters, and no access to the sea combine to produce high mineral content in the Eastern Rift lakes. Because the Rift lies along parallel fault lines, it has always been a very active volcano area. Mount Kilimanjaro in Tanzania is part of the Rift Valley system.

This 'unbalanced' environment with its clash of forces also happens to be fertile ground for life and its development. That will be one of our major themes—'good stuff' occurs around rifts as forces work to bridge the divide and restore balance.

Ethiopia has been called the 'water tower' of east Africa because of the many rivers created by its highlands. In southwestern Ethiopia, the Omo River, which shares a drainage divide with the Nile, is one of those. The Omo travels about 200 miles through a steep-walled valley before it runs into the northern part of Lake Turkana, most of which lies in Kenya. The Omo National Park is the largest park in Ethiopia. Wildlife abounds including elands, buffalo, elephants, giraffes, cheetahs, lions, leopards, zebras, monkeys, hippos, over 300 species of birds, and many other animals. Today it is a tremendous expanse of wilderness, but it appears that for millions of years it was an important incubator for the development of various human species. And you can see why—rich soils provided in part by many volcanoes, water availability because of the many lakes and rivers, a temperate climate due to higher elevations though situated at the equator, and an incredible abundance of wildlife and food.

It was on opposite sides of the Omo River where anthropologist Richard Leakey discovered the bones of two humans in 1967. They have recently been reexamined, and now the bones of each are dated at about 195,000 years old, the oldest modern human bones ever found. Omo I appeared to have modern features while Omo II appeared more primitive. We mentioned previously that new species generally begin in small groups on the periphery of the existing population. Did it happen here? Possibly.

DNA Studies

There is very little genetic diversity among human beings, no matter who they are or where they live. Two chimpanzees sitting next to each other have more genetic diversity than do an American and a Chinese living on opposite sides of the world. That is because chimpanzees have been around a lot longer than we have. All things considered, we are recent newcomers.

There are several approaches that researchers use in their efforts to try to understand the social and cultural development of human beings and their movements around the world: archaeology (artifacts mostly), paleontology (bones mostly), and more recently, DNA studies. Surprisingly, all of the bones found around the world from early humans would fit in the back of a pickup truck. That is not much to go on.

DNA studies suggest that the San of Africa, often called the Bushmen, are the oldest modern humans, and the San themselves say they are the 'first people'. A recent study by a team led by Sarah Tishkoff from the University of Pennsylvania found that modern human migration is believed to have emanated from the coastal area of southwestern Africa near today's border of Namibia and Angola. Not surprisingly, this is an indigenous San area. Most of the surviving San people today, numbering between 50,000 to 100,000, live in southwestern Africa in the Kalahari Desert, which is located in parts of Namibia, Botswana, and South Africa. The Sandawe are an old San tribe still living in Tanzania.

The Sandawe and other San tribes that still survive, even after tens of thousands of years living apart, continue to exhibit almost identical behavior. We see it also in the San and another tribe, the Hadzabe, which split off from the San probably 70,000-100,000 years ago. Genetically, the Hadzabe people appear relatively distant from the San, due to the length of time living apart, but behaviorally very close. With these two very old groups both using a language based on clicks and tonal expressions, illustrated by the ! in !Kung and the / in Ju/'hoansi, it demonstrates that the click-language may have been the beginning of modern human language.

The click-language in use by a San group today has about 141 different speech sounds, or phonemes, whereas English has about 40. The only known non-African group to use a click language was an Australian Aboriginal group, another very old group. You will soon discover why that makes perfect sense. Incidentally, the clicks may have initially been used as communication tools among hunters, and then language developed from that. The click sounds make up about 40% of the San language.

This similarity in behavior is the same thing that author and biologist Stephen Gould saw in other species; species can go through long periods with very little change and can therefore become very homogeneous. Consequently, a stasis develops. If a human population remains large enough to maintain homogeneity and stays fairly isolated, behavioral and other traits can go on for tens of thousands of years as they have with the San and the Hadzabe. But if the climate changes dramatically, then a modification of the species may occur, unless the species is very adaptable. And that adaptability is what modern humans seem to have.

A lesson should be drawn from the Hadzabe-San example. Though genetically different, relatively speaking, their behavior is very similar. DNA studies can lead us to see differences among people which behaviorally might not exist. As noted earlier, humans are vastly more alike than different. The differences genetically are miniscule.

Climate has been a consistently strong variable in the development of human beings and everything else on earth. The earth seems to have an ice age about every 100,000 years, or has during the last two million years or so. Partly this is due to a slight wobble in the earth's rotation on its axis that takes about 25,700 years to complete. Also, the tilt of the earth's axis moves between 22.5 degrees and 24.5 degrees every 41,000 years. And then there is earth's orbit around the sun which goes from elliptical to circular every 100,000 years. Plus, volcanoes and the earth's tectonics are always changing the surface of the earth, sometimes rearranging the flow of water and air currents. The last 10,000 years have been mostly stable climate-wise, and have allowed modern humans to develop as never before.

About 200,000 years ago, another cooling period set in, lasting 70,000-80,000 years. The drought conditions that arose in Africa, however, were probably the catalyst needed to facilitate the emergence of homo sapiens, followed thousands of years later by homo sapiens sapiens. As I have stated before, stars form mostly around turbulence. It is the same with everything else—no imbalance, no change.

I mentioned that the modern human species, including the homo sapiens sapiens subspecies, probably formed as other species do—out on the periphery, isolated from the ancestral group. If the San were in fact one of the early groups of homo sapiens sapiens, we would expect to see characteristics in them similar to an isolated species-forming group. And that is exactly what we see in their behavior—semi-isolated, extremely resourceful, strong group-orientation, and cooperative behavior, just what would be needed if a small group were to survive on the periphery.

Severe conditions challenging survival would, through natural selection and genetic drift, produce specific individual and group characteristics that would be advantageous for dealing with the harsh environment. Genetic drift occurs in small populations in conjunction with natural selection. A certain randomness of gene selection can occur which can determine the characteristics of future generations. Genetic drift is not always a positive. But if significant genetic drift was at play in the early San, we lucked out because as a prototypical blueprint for the future, it is hard to see how we could have done much better. Everything positive about modern humans is based on the San—the negatives came later.

The San may have been the 'first modern humans', but you will see that the San were never what some people might call primitive. We have discovered that bacteria are complicated; the first modern humans were no different. In an ironic twist, we will discover that the basic human and gender rights we battle for on a daily basis in the modern world have been the foundation of San society for over 100,000 years. After you learn more about the San, you might be tempted to question your assumptions about what the word civilized means, though I suspect that many of you already do.

Who Are the San?

The San themselves do not really have a collective name for all of their many distinct tribes. The term San, meaning outsider, was given to them by their genetic cousins, the Khoikhoi, pastoralists who broke away from the San about 2,500 years ago. Anthropologists often combine the two groups and call them the KhoiSan. We will focus our attention on the San. They preceded the Khoikhoi by over 100,000 years.

The terms San and Bushmen are often deemed to be derogatory by the people referred to as the San. I am not aware of another name that would be acceptable, so with apologies and without any intention of disrespect, I will regretfully continue to use the term San. But you will see clearly that I have nothing but the utmost respect and admiration for them.

Recent history has not been kind to these first modern humans. They have repeatedly been forced by governments and invaders to leave the hunting and gathering areas where they have lived for tens of thousands of years. When Europeans started settling in Africa, they hunted the San like wild animals, almost to extinction. Plus they were tortured, harassed, kidnapped, raped, enslaved, imprisoned, and used as fodder in the Europeans' wars.

White Christian settlers debated whether the San were really human and if the Biblical scriptures applied to them—then slaughtered them by the thousands. Consequently, their population has fallen from several million to 50,000 to 100,000 today. It is not treatment that should be borne by anyone, let alone the world's oldest living human group, and the people who ensured survival for this species. It shows clearly the ugly human elements that have managed to solidify a spot for themselves over the past few thousand years.

The European-San interaction reveals the underlying process that we have been discussing. You will recall that cyanobacteria enabled the creation of other life forms which then fed upon cyanobacteria. The structure, whatever it is, captures the creative-formative-productive energy (CFPE) and uses it for its survival needs. Here we have the Europeans with significant amounts of IMPACTS energy of their own. If not, they would not have been able to emigrate in such large numbers and build a society from scratch. But these IMPACTS-laden people are using their IMPACTS energy against the original IMPACTS! This is a vivid demonstration of how the real world works—and doesn't work.

European societies had developed a nucleated structure with strong SN characteristics. The original IMPACTS, the San, had remained a non-nucleated group. Therefore, they were easy pickings for extreme male behavior. Please make a note of what the extreme male Europeans were doing—they were accreting lands that did not belong to them, and they were doing it violently. Accretion and violence, generally emanating from the male structural-nucleus, work in tandem. This has been a recurring theme for the past few thousand years, and continues to be a prominent part of the world culture.

All things being equal, the character of the structure really depends on the quality of the SN elements. If the SN leans toward benevolence, that is what you get. If it leans toward belligerence, that too is what you get. With extensive IMPACTS energy available, anything is possible, good and bad. You can get manned flight to the moon, and you can get Hitler, Stalin, and a host of others, including present-day versions. The murderer on the street will often go to prison. Heads of government, whose victims can number into the thousands and millions, usually find ways to avoid prison and the tag of murderer, though their crimes can be egregious. It is important to understand that the SN is the entity that is defining right and wrong, friend and foe, murder or self-defense. There is one standard for the guy on the street and another one for the halls of power, as all of us already know.

Even after what they have been through, the San have continued to cooperate with those who have studied them. They want the world to know their story, and it is a great, yet tragic, story. But it is also the same story repeated throughout nature and seemingly throughout the universe. It is the tale of two different kinds of energy—one connecting, sharing, creating, and producing, the other taking and dividing.

It is believed that the San basically had the free run of much of central and southern Africa for tens of thousands of years. The San included many different groups and tribes; for example, Sandawe, !Kung (also known as Ju/'hoansi), /Xam, and Hai//om, to name a few. All San tribes have always been very independent and largely self-sufficient, living nomadically in semi-isolation and often having little to do even with other San tribes. The contemporary world is built largely on the model of the independent San tribes, except that today an SN structure is running each 'tribe' or country.

The San were a hunter-gatherer society, with 80% of the food gathered by women—edible plants and roots, fruits, nuts, berries, and small animals providing most of the protein for the group. What was gathered was usually shared only with the immediate family, while meat was shared with the group. Gathering provided women with a social platform that solidified relationships. During such activities, they might also discover animal tracks and other secrets of the bush, including new water locations. Everyone worked together and shared knowledge.

San Family Life and Gender Roles

Women were very autonomous and were treated as equals. But then, why would they not be? The San would not have understood a concept of inequality just as they did not understand what property meant. Later, their genetic cousins, the Khoikhoi, and Europeans too, would accuse them of stealing cattle. But to the San nobody owned cattle or the land on which cattle grazed. Now you can clearly see why many IMPACTS in today's world care little for possessions and property. Such attitudes emanate from our ancestors, the San.

Men and women alike had extensive knowledge of plant life, including those plants with medicinal properties. Men hunted the bigger game, antelope being the most targeted. Meat was distributed to the other members of the group by the 'owner' of the successful arrow that killed the animal. That arrow could be owned by men or women. Men also produced clothing from animal hides and crafted various tools, weaponry for hunting, and musical instruments.

For these early San hunter-gatherers, life could be precarious, but it could also be joyous and festive. They learned how to survive and enjoy life at the same time. Some have called it the first affluent society. If a large animal was killed, other bands might be invited to share in the bounty since there was no way to preserve the meat. Such behavior increased goodwill which might be needed in difficult times ahead. But as much as anything, it appears that sharing and cooperating came naturally for the San.

The nuclear family was the primary social unit. Hunter-gatherer bands would be formed of different families, each band comprising 20 to 50 people. Some of these mobile groups followed game, water, and food around the countryside, living in harmony with nature and the seasons, having no domesticated animals or crops and only minimal possessions. Other groups lived along the coast, dining on seals, shellfish, crayfish, birds, and even an occasional beached whale.

There was no hierarchy and no leadership position based on heredity. The male head of the main family or another elderly man with respected qualities would most likely lead the discussions where a group decision was required. Lengthy discussions would accompany the resolution of disputes. Everyone would be encouraged to offer their thoughts and opinions until a consensus was reached. True democracy was a way of life. If a consensus could not be reached, the families would simply divide up and go their separate ways. Among San tribes, there was no formal military system; there was no need for one with San generally being pacifists.

Male behavior extolled in today's society would not have been tolerated in San society. Aggressive competition, one-upmanship, self-aggrandizement, or any other display of non-cooperative behavior would have been summarily discouraged. Alpha-like behavior would have been a threat to harmony and therefore to survival. Even with the kill of a large animal, the hunter's achievement was downplayed while sharing was promoted.

Gender roles were minimal, but women did assume the principal duties of child-rearing and cooking. But those responsibilities were coupled with the major decision-making that developed around the children and their eventual marriages.

Children were adored by the entire San tribe and lavished with love and attention. San fathers were affectionate and devoted, though most of the children's time was spent with the mother. Relationships between parents and children were usually emotionally and physically close, and non-authoritarian.

A look at a present-day !Kung mother indicates the influence and devotion of the mother in San culture. A !Kung mother generally nurses the baby until the age of four and sometimes to age six. Such suckling induces hormonal changes that inhibit ovulation and hence pregnancy. Dietary stress may have something to do with it as well. Carrying the baby in a sling, he/she suckles almost at will throughout the day during the first two or three years, creating of course a strong bond between mother and child.

The young children of the !Kung accompany their mothers wherever they go; they will travel thousands of miles together in the first few years. But as tough and resourceful and nurturing as they are, if they have the average of five children, generally they can expect only two to survive to adulthood, marry, and have children. Can you imagine what it was like when the climate was even more challenging? That is why it took tens of thousands of years for the population to grow appreciably.

Public life and domestic life were basically the same for everyone in the San tribe. There was no place to hide in social relations. The structure of the group was built on egalitarianism and cooperation, and those aspects were promoted by the openness of the living situation.

The San were efficient in their utilization of the environment. For example, they would bury ostrich egg shells filled with water for later use, drink the stomach juices of freshly killed animals, and to prevent dehydration, smear their skin with animal fat. The trick was to not overlook any aspects of the environment that might aid survival.

The San Community

San community life and concept of society can be symbolized by a circle. A circle has no beginning and no end, no gaps, no top, no bottom. That was their philosophy. Their living environment reflected that circularity. They would build their huts in a circular fashion around a large communal area, the door of each hut facing the center of the communal area. A campfire would burn in front of each hut. All activities of the group (except sleeping and intimacy) would take place in this communal area: storytelling, children playing, dancing, singing, healing ceremonies and rituals, and cooking. At night, families might move from campfire to campfire, socializing and exchanging gifts, always showing respect for one another and exhibiting a certain dignity. This was the foundation of modern humans—strong human bonding. These were our beginnings.

Storytelling by elders preserved San history, traditions, and wisdom. In many of the stories of the Sandawe, the San tribe of Tanzania, San identify with small animals which have to use their cunning and intelligence to defeat their larger and more powerful enemies. The !Kung of the Kalahari region sometimes refer to non-San people as 'animals without hooves', meaning they are as dangerous as predator animals.

Cave spirits occupied a central place in Sandawe spiritual life along with ancestor worship and divination. The Sandawe regularly made sacrifices to cave spirits living in the hills, and were very careful not to hunt or gather wood near caves. They also sacrificed at the graves of their ancestors in order to maintain good relations—and bonds—with the spirits of the deceased. Ancestor worship and caves would play prominent roles as people spread out around the world.

Bands would have remained for long periods in particular environments if food was plentiful and conditions were good. But during lean times, they would often split apart and join other groups. That meant getting along well and cooperating with other people.

One way this was accomplished was by setting up what was called hxaro exchange between neighboring groups. One person in one group would be the hxaro partner of someone in another group. Gifts between the two would be exchanged, although the gifts would have no relation to survival needs or economic needs. The purpose was to form a relationship; what was exchanged was of no consequence. In the future if there were hard times, the social networks were in place that could be utilized for helping people through the crisis.

Steatopygia

Even the physical appearance of the San is particularly unique among African tribes. They are of short height, light-skinned (copper-brown), and have tightly-coiled 'peppercorn' hair, high cheekbones, and epicanthic eye folds commonly seen among East Asians. San facial features are seen around the world, from Asians to Eskimos to Latin Americans to Caucasian Americans to Europeans. That should tell us something very important if we will stop to think about it for a second. Major insights are right in front of us. All we have to do is look at them and think about them—open-mindedly. The reason that San features are found around the world is because we are all derived from the San. It is not a mystery. The force of the SN hierarchal structure prevents us from seeing the obvious because it distorts our picture of reality. It is the human version of Einstein's general theory of relativity where mass and energy distort space-time.

Another feature seen prominently in the past among San women, and to a lesser degree among San men, is a condition known as steatopygia, though it was probably prominent in both sexes among early modern humans. The body stores fat in the thighs, buttocks, and midsection so it can be used when times are lean, leaving the arms very skinny. This would have been an indispensable aid in the survival of modern humans in Africa given the harsh environmental conditions, and there were many such conditions.

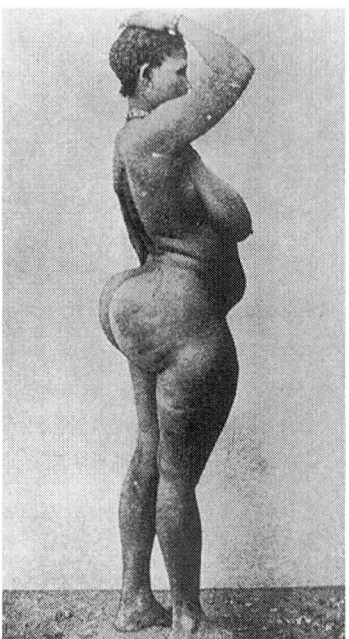

Public Domain Photograph of Woman with Steatopygia

Steatopygia has not been observed among farming people. It is primarily seen today in two living populations: the San of Africa and the Andamanese of the Andaman Islands off the eastern coast of India. Plus these two groups share the same peppercorn hair. However, the Andamanese are very dark, while the San are light-skinned. The Tasmanians, who formerly lived off the southern coast of Australia until they were exterminated by the British in the mid 1800s, had the same physical characteristics as the Andamanese, including the peppercorn hair and dark skin.

Many of course will say that the two groups were not San-derived because the color of the skin was much darker. But that is simply because we do not know all there is to know about skin color or anything else. Common sense should prevail when given the opportunity. This is a case in point. You will note that the two

San-like groups above were both living on small islands. That was common as humans spread out around the world.

With the disruption of their traditional lifestyle, the prevalence of steatopygia among the San has declined. Access to food is not as unpredictable as it was in previous times. Hence, there is less need for the body to store so much extra fat.

The Shaman

At the center of the San community was the shaman. If you understand the San and the role of the shaman in the San's everyday life, you will understand how the IMPACTS personality profile developed and how and why it has been both the fuel and the engine that have driven human society forward since the beginnings of the San.

Who was the shaman and how did he/she come upon the scene? The first shaman-like person may have been the progenitor of homo sapiens about 200,000 years ago, though this would not have been a San-person. The San probably developed 120,000 to 130,000 years ago or so. The biggest difference between homo sapiens and previous human species is innovation, which I believe came in with the shaman. There is no evidence of a shaman role in previous human species.

As we have seen, creative-formative-productive energy (CFPE and IMPACTS energy) helps develop the structure and then serves as the innovative-maintenance-improvement-change energy when the nucleus emerges and captures it (the CFPE). At this stage in the process of modern human development more than 100,000 years ago, we have a 'prokaryotic cell' with a loosely-defined nucleoid and a circular configuration of its DNA. The shaman is the DNA, and everything revolves around him or her. The shaman is carrying the information needed for ongoing duplication and survival of the tribe just as DNA is for cellular and life duplication. The roles are exactly the same and they are both 'new beginnings' and foundational: DNA for life itself, and the shaman for the new human species. The shaman is also the nucleoid, and he directed the 'prokaryotic' San tribe in a 'loosely-defined' manner.

**Everything is Connected in the San Tribe. It is an
Uninterrupted Feedback Loop.**

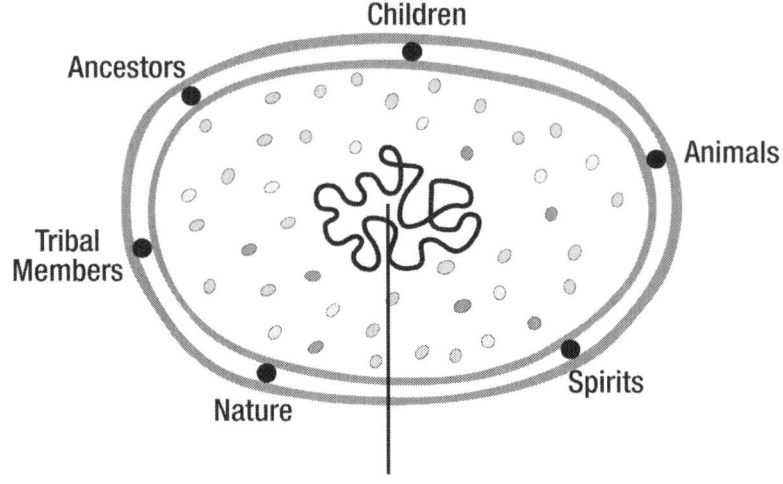

Shaman is the DNA
of the San tribe.

**The 'Prokaryotic' San Tribe and its Shamans Can Be
Thought of as a Node of Energy Production.**

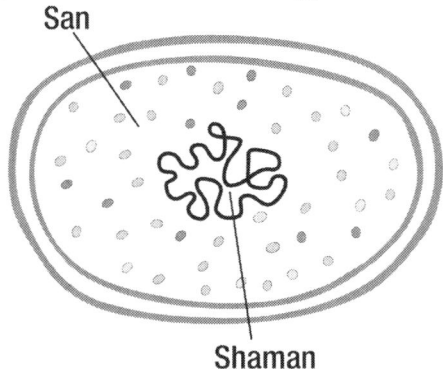

Bacteria ruled the earth for 2 billion years or so, laying the groundwork for eukaryotic life forms. The same happened with the San and shaman blueprint—they and their San-like descendants 'ruled' basically by themselves for tens of thousands of years, until the nucleated structure developed alongside agriculture. Then, just as happened with bacteria, the San-like descendants, the IMPACTS, were brought inside to provide the primary energy for the structure, both creative-formative-productive and innovative-maintenance-improvement-change.

'Eukaryotic' Civilization Developed After Agriculture

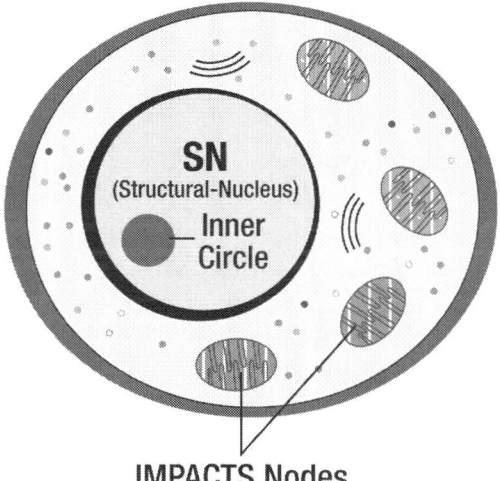

IMPACTS and IMPACTS Nodes
Became Civilization's Mitochondria

You will note that an atom does not exists until the proton captures the electron; in other words, until the electron becomes 'semi-stationary'. Prior to the formation of the atom, a proton is just a proton. With the capture of the electron, it becomes the nucleus of a structure with workable energy in tow. In our brief comments about a wolf pack in the previous chapter, you will recall that the abused and lowly omega can run off, find a mate, and become an alpha. Once he unites with duplicating energy, everything changes for him as he then possesses the power to change his environment.

With the development of agriculture, the valence-electron-IMPACTS of human society, the San-like descendants, were captured and became semi-stationary. Thus, a nucleus could begin forming, and it has been strengthening ever since.

Aggressive alpha behavior is seen among some of our primate relatives such as the chimpanzee, but in others there is less male aggression and dominance and more female behavior. Gorillas have a close-knit family life and males rarely kill though they will fight over females—but human beings do that. We are also as close genetically to the peaceful bonobos as we are to chimpanzees, and bonobos have never been known to kill one of their own kind.

The recent discovery in Ethiopia of "Ardi", a 4.4 million-year-old hominid skeleton of the species Ardipithecus ramidus, suggests that these early humans might possibly have exhibited little alpha behavior as well. Their canine teeth were found to be smaller and less sharp than those of chimpanzees, which chimps use for violence, and it appears that both sexes had the same-sized teeth, suggestive of more equality. It looks like humans and chimps split off from the same ancestor 5-7 million years ago.

It would not be surprising to find that violence predominantly emerged in humans only with the development of agriculture 10,000 years ago and not millions of years earlier as previously believed. As we have discussed, formative energy is mostly female and male energy usually comes along after the development of the structure is well underway. The structure in this case is the modern human.

In times of deep stress, nature will sometimes provide a virgin birth. It has been observed in sharks, lizards, and other animals. This allows the species to continue, though of course it is a clone of the mother and therefore does not have any additional genetic diversity, which at times is needed to adjust to a changing environment. It is a stopgap measure just to keep things going.

That is what was needed at the origins of modern humans—a virgin birth of sorts, a genetic community awash in female creative-formative-productive energy without the male issues of dominance and aggression. And that is what it was. It may have started with the beginnings of homo sapiens 200,000 years ago, and then possibly increased with the development of homo sapiens sapiens as the San or San-like people emerged, or it may have started long before with prior human species. Should it be surprising then that the San are so clone-like? After 100,000 years, their behavior is basically unchanged. But you might say, "I thought innovation was an important part of the San-shaman profile. How does that equate?"

For the San to have survived for such an extended period shows immense innovation. You don't have to invent a machine in order to be innovative. Sometimes just finding ways to survive is enough. San-shaman innovation started from the intense desire to help others—to keep bonds from dispersing—to leave no one behind. All other forms of innovation sprang from those goals. That is why so many of today's IMPACTS are in health-related professions, and why so many great inventors and discoverers stress that their work should be used for peaceful purposes and to help all human beings.

I believe that the first shaman 200,000 years ago may have been from the autistic spectrum, possibly an Asperger syndrome-type profile. Of course during this period there were no categories of normal and abnormal behavior, and no spectrums of any kind. All of that is a fairly recent invention with the number of categories growing by the day. But as we will continue to see, that is the way a hierarchal structure operates—it keeps dividing people instead of looking for commonalities. It is exclusive and not inclusive. During the beginnings of homo sapiens, there would have been no stereotypes and no need for ostracism or shame. The 'imbalance' in the shaman may have been what was needed to deal with the imbalance in the environment at that particular time.

The beyond-the-norm passion and nurturing attributes, and the pulling for the underdog, could actually have developed within the shaman from the need to deal with the difficulties and sufferings emanating from autistic family and community members. A scientist on a recent History Channel show suggested that the mutation that started our modern human species probably occurred in a male. He did not say why he thought so. If it were an 'autistic' mutation of some kind, that would make sense in that it would be a way to 'feminize' the male half of the species, stripping away much of the masculinity. It would also explain why autism is found almost exclusively in males. The shaman profile may also have been more prone to mutations, and thus extremely valuable in an additional way. Tremendous advancements can originate with mutations.

Contrary to what the public usually thinks, distinct advantages for the human race emanate from the autistic spectrum, including from the extraordinarily inventive Asperger's part. The autistic scale would have had no advantages for humans among females, but it would in males as it would have provided a 'carrier gene' for inventiveness in males that had not existed previously. Males and females could then both 'create' and bring things 'into being'.

Essentially, it appears that males exhibiting autistic-spectrum traits are detached from certain aspects of typical maleness or left-brain emphasis. We see this same lack of maleness in varying degrees in IMPACTS men, by and large. That is one reason I believe that IMPACTS males exist in proximity to and sometimes overlap with the autistic-spectrum. Another is their natural innovation.

The San-shaman was the creative-formative-productive energy of the San tribe and the innovative-maintenance-improvement-change energy as well. And let's not forget the strong protective element that comes with CFPE. So innovation was housed permanently within the shaman, to be called upon as needed, and often that was during turbulence. As we have seen with star formation, turbulence often ignites 'new beginnings'. That is what creative-formative-productive energy (IMPACTS energy) is all about—new beginnings. We see it in the valence electron bonding with other electron energy to produce a molecule, in molecular hydrogen gas clouds where stars are born, in bacteria that laid the foundation for eukaryotes, in San-shamans who searched for solutions, and in today's IMPACTS who innovate and discover.

Seen within this context, the Big Bang was an IMPACTS event—a new-beginnings event that resulted from turbulence of some sort. Comets and meteors are new-beginnings energy too. They reside on the periphery and, as we have noted, contain organic molecules and water—the new beginnings of life. As we will continue to see, new beginnings usually emanate from the periphery.

The value of the shaman to the tribe was more than a search for answers; after all, anyone can search. It was the passion and commitment that the shaman possessed to find the answers, and this existed around the clock. The passionate search was a catalyst for a more expansive view of the world and reality, a view that was constantly evolving because of the efforts of the shamans. Most importantly, the human environment revolved around the shamans and consequently demanded innovative, caring, searching, shaman genes.

It is often called the 'shamanic sickness', an originating challenge that led many to become shamans. Shamans and would-be shamans were undoubtedly more sensitive to their environments than were others (more aware), so therefore their reactions to events would be different, illness often being the result. This sickness could have been caused by the death of a family member, an injury from a wild animal, the rejection of a lover, a congenital condition—anything. Sometimes this sickness was so harrowing that the would-be shaman felt that his mission in life was to recover from the illness and then to save others from the same experience. He had been 'chosen' to help others by events beyond his control.

Tribal members were very much a part of the shamans' work and efforts. The early San tribes had so much of this searching-solution energy that often one-fourth to one-half of the entire tribe would be shamans—men and women—but usually more men. Sometimes the solutions would be found, sometimes not. But the problem-solving mechanism was in place, and everything was built around that. When the SN came along after agriculture, it would do the same thing—build itself around the problem-solvers. But the SN would give them significantly less freedom.

In many shamanic societies, if a young person showed special abilities such as predictions through dreams or more sensitivities to others and nature, or displayed even small amounts of psychic or artistic ability, he or she would often be chosen for training under the tutelage of a shaman. Even those who had physical problems such as migraines, epilepsy, and skin disorders, were often considered to be future shamans. Why? Probably because shamanic tribes had learned that 'imperfections' might be masking other abilities, that dealing with adversity and turbulence can often be an attribute for a person and the community.

Many of these young people undoubtedly leaned towards the autistic side, a situation that has a negative inference today. But that is because our modern world is distinctly left-brained, and its SN leaders are looking for a homogeneous following. They are not generally inclined to place value on those who are not necessarily contributing to that homogeneity. Plus those on the autistic scale may not be contributing to the economy, which gives the SN its strength.

If you were in control of 100 people and they were working on an auto assembly line, would you prefer that they be mostly alike or very different? That is our world today—a production machine. If we are all alike, it makes it a whole lot easier on the SN.

On our straight line graph from Chapter 3 signifying the engaged-autistic spectrum with the left end being engaged with the environment and the right end disengaged, the San-shaman would of course have been somewhere on the right side. This follows the model of the human brain also, where the right hemisphere is the creative and random, yet holistic side. Some shamans would have been further to the right than others.

Engaged with the external environment   Autistic or not engaged
shaman

This position placed the shaman basically between two energy fields—everyday reality and a 'spiritual' reality where he could go to search for answers to the problems of individuals and the community. Transformative ideas did not come from the mind that was tightly engaged with regular reality; they came from 'out there'—the spiritual reality. This will be the recurring model as human development goes forward, right up until today: the discoverers will often have only a loose connection with the everyday world. That is one reason Asperger-profiles can be so inventive—their energy is not being 'wasted' on connections with everyday reality and its endless demands.

The spirit realm has been passed down through the ages, and humans still often think of the really important answers as coming from a vague 'spiritual' reality. But there is no evidence that any such realm exists except in one's imagination, which is exactly where it existed with the shaman. Spirituality has become the abode of the unknown.

We mentioned that the left side of the brain is usually the masculine side and that the San were mostly female-oriented with a clear diminishment of masculinity. That is why we see so many IMPACTS men displaying female qualities, and why so many of the great male innovators and thinkers through the ages, including today, are clearly androgynous. They are disconnected from much of their masculine side and therefore have more freedom to think creatively. Creative, new-beginnings energy is female energy.

Think about Gandhi, Buddha, Martin Luther King, Jr., and other peacemakers through the ages. What were they all espousing in one form or another? Giving up the male ego or stereotypical male behavior—in other words, going all the way back to the roots of modern humans which just happen to be the San tribe and specifically the San-shaman. Even Jesus and other religious leaders were essentially San-shamans at the core. Almost everything in the modern world emanates from the San tribe, ultimately from the San-shaman.

Isaac Newton was only tenuously connected with the everyday world. When he made his most important discoveries about the universe at ages 22-24, he was on a two-year break from the university because of a plague epidemic. But isolation was a way of life for Newton. His personal relationships were few and far between. We now know that Newton was just as interested, or more, in alchemy as he was in science. Most discoverers are not as estranged from everyday life as was Newton, but they are usually on the periphery in one way or another.

It makes sense when you think about it. Change comes to the atom from the periphery because the energy of the nucleus has a tight control over the electron energy closest to it. It is the same with a powerful human organization or human paradigm; the strong pull of the core produces a homogeneity that can cause change to be extremely difficult from within and without. Therefore, change energy usually has to connect with like-energy on the periphery if the structure is to change. By the way, so estranged was Newton that he did not even share his discoveries with the world until 20 years after he made them.

The analogy between the shaman and the valence electron is again very relevant. The valence electron is semi-engaged with the nucleus. It has the freedom and 'motivation' to look for a connection with other electron energy—therefore, restoring balance. It 'steps outside' of the energy field in which it is semi-captured and produces a solution—a molecule. The San-shaman has the same freedom and motivation to find solutions to the imbalances in his environment, and he too will step outside of the predominant energy field, everyday reality, and return with solutions.

You can see that the developing civilization of the San and shaman is following the model of the physical world. That is why modern humans have 'progressed' as rapidly as they have. The structure of modern human civilization is in sync with the

149

universal model, though it is obviously not an exact calibration. Previous human species had no 'valence electron', just protons and electrons, males and females. Now there was a combination model, an androgynous model.

Philosophy and Religion

The whole philosophy of the San-shaman was built around these questions in the shaman's mind: What needs to be done at this critical moment in order to optimize the potential of the situation, including the short-term and long-term? What can be done to make the best of a not-so-great situation? How can I bring wellness? If the valence electron could 'think', it would probably be asking itself the same questions. But then, our thinking may be a poor imitation of electron 'awareness and calculation' across the universe. As noted earlier, we don't have a clue how entanglement works among subatomic particles; the particles theoretically know the physical 'states' of their entangled partners even when they are trillions of miles apart. So our thinking is probably very much on an elementary level as it compares with communication occurring throughout the universe.

The San and shaman ascribed 'humanness' to everything, or at least that is the way we would phrase it. The mostly right-brained San tribe believed that everything was alive and interconnected— trees, rocks, people, animals, water, dirt, clouds, stars, the sun, and everything else, and that all these things had spirits which were active. With our increasing knowledge of how everything works, including the recent discovery that snowflakes and raindrops often form around bacteria, their beliefs do not seem so far-fetched. Actually, they appear to be more realistic than do ours. But this is the way creative-formative-productive energy works before a nucleus forms and chops it up into little pieces—CFPE is inclusive and sees connections. It is anti-dispersal.

Our view is pretty much the opposite of the San's. Ours is atomized, separated, disconnected, and divided. The wholeness is hard for us to see. But again, that is the way a left-brained hierarchal structure works. The flow of energy to the top makes it difficult for the components to see and experience cohesion.

If everything around you were alive and made up of spirits, and you were then able to please those spirits through good deeds

150

and sacrifices, you had a fighting chance of living harmoniously with the environment, or so it was believed. You can see how the spirits could easily morph into gods and goddesses. The sun spirit could become a sun god, and just as you had to please the sun spirit, you would have to please the sun god.

It is clear how religions would develop in which you would be judged ultimately by the good deeds and actions you performed during your time on earth. Such thinking was not that far removed from the San, but the focus was entirely different. The San were concerned with the well-being and survival of the tribe or community with little emphasis on the individual. Religions were mostly concerned with the well-being of the soul of the individual. But that was just a reflection of the gradual shift in the structure of humanity from a circular environment to a hierarchal one.

SN development meant that individuals would become more and more responsible for their own lives, the hierarchal structure preventing them from looking to one another for 'salvation'. They would have to find those answers, and that help, elsewhere. Often, it was just 'me and my God', and still is today, as the structure of the SN world continues to disperse people one from another.

After one God replaced the many gods and goddesses, SNs would often demand, "Believe in our God and convert to HIM or else." This tactic in the hierarchal war on cohesiveness and community was utilized frequently against paganism, which was at heart a female 'religion' centered on the earth. Paganism was essentially a holdover from the San, as was 'Goddess worship', which was most likely based on the San reverence for women. Agriculture was the dividing line. Everything before was basically San-like and female, and everything afterward was SN-like and male with the San-like female elements captured.

You can see that religion as we have come to know it is pretty much the opposite of paganism and Goddess worship just as the SN is the opposite of the San. Again, it is male versus female. Paganism has a negative connotation today, a result of being demonized for thousands of years by the SN. When the SN doesn't like something, it demonizes it rather than trying to work with it. It is part of the SN philosophy of 'this is my world'. We can see that attitude clearly on the world stage.

The SN took the earth-centered 'religion' and stuck it in the sky—'out there' somewhere, far away from the earth and people, making it easier for the SN to exploit both. If the 'spirits' were in everything around you, you were never separated from them. If they were in the sky, they were harder to reach. You can see why Jesus and others said, "God is everywhere—always with you." He was talking to the 'underdogs' of society who were receiving little in aid from the SN leadership. The attempt was to bring God back into people's lives where SHE had been originally, before the SN had arrived and changed her to a HE.

If you are at the top of a hierarchal structure, you do not want a bunch of gods and goddesses diverting the attention, energy, and loyalty of your subjects. You want as much of that as possible flowing in a stream in your direction. How better to do so than to actually proclaim that you are part-god or the offspring of a god or gods, which many 'leaders' did.

Protectors

The San and shaman placed themselves in a subservient position to the power they saw all around them. They were humbled by their surroundings. The present-day world has done the opposite—the controlling-SN has put nature in a subservient position along with the rest of us. The serving-others attitude of the San and shaman would be captured and abused by the nucleated forces that would arise after the San and shaman and their San-like descendants had laid the solid foundation for modern humans.

Interestingly, the San tribe's 'reliance' on the shaman appears to have created an ongoing need in much of humanity for a force that could be called upon to help in critical situations—an omnipresent savior, protector, and healer. As the development of human beings continued, religions would emerge that would try to satisfy these needs, and SN 'leaders' would profess their ability to do the same. Religious and SN leaders were just trying to take the place of the shaman and wield his power at the same time. But generally, they could not provide the essential elements that made the shaman so valuable: proximity, availability, ongoing innovation, personal commitment, an expansive view, and genuine concern and caring—person to person. The attempts to step in and re-create the shaman-community trusted relationship would invariably meet with only limited success.

These saviors and their religions started emerging after the development of agriculture, after the structural-nucleus (SN) of society had started taking control. Human beings were relegated to unequal status as the hierarchal structure broke cohesion among people and accreted their energy, directing it away from one another. In the San civilization, everyone had been equal, and people had relied on each other for help and sustenance. In the new SN world, people were not equal, and they were being continually divided from others. The world had been turned upside down. That is why 'saviors' were needed and welcomed by everyday people, but not necessarily by the SN; e.g., the story of Jesus.

A good example of what has happened around the world is seen in the Kalahari Desert today, where many San have been relegated to basic serfdom on farms. Shamans have become itinerant, traveling from farm to farm to perform rituals and maintain a sense of community—trying to keep the bonds together and prevent dispersal. But as the San civilization unravels due to modern world encroachment, fewer shamans are found because the new structural environment will not tolerate any remnants of the previous power structure. The San community revolved around the shamans; it adopted the beliefs and insights and views of the shamans; shamans carried tremendous influence. As the San society has become threatened, shamans have become protectors 'politically' as they encounter 'non-egalitarian' conditions on their travels.[1] In other words, injustice has reared its ugly head, and the shamans are attempting to combat it.

What they are trying to combat is the accretion of energy by the SN in tandem with its general disregard for the communal standards that existed before agriculture. But the shamans will be defeated as the hierarchal structure will methodically squeeze the protective-communal part of their energy from the 'fields' and send it in other directions. If there is value from their energy that can be utilized by the SN, it will be captured. If it gets in the way, it will be destroyed. After all, the SN has decreed itself as the protector of everything and everyone within its own self-defined community. Outside agitators are not welcome.

This has been a consistent problem for the SN—how to deal with these protective-communal shamans, or IMPACTS, who also possess many other abilities badly needed by the SN. The solution for the SN has been to reward, through natural selection, the skill set desired and to get rid of or ignore the people who do not possess the 'correct' traits and attitudes.

Currently, most of the political leaders are looking out for the SN and not the people. Actually 'the people' has become a pejorative term as the SN again has demonized anything that smacks of cohesion and tight 'people-bonding'. Why do you think the Church of the Middle Ages was so opposed to pagan practices, 'witches', and even midwives? Because they were part of the cohesive, female elements of society that the hierarchal structure wanted to dissolve. They were impediments to the flow of energy. They were anti-dispersal bonding agents.

Control the Energy

The SN in power wants to control the character of the bonding within its domain. Let's look at a hydrogen atom again with its valence electron. When the valence electron bonds with another valence electron, the nuclei of course are part of the new structure and therefore influence its physical characteristics. Whereas it was the electron that had been captured originally, now it is the bonding that is captured.

It is the same in the human world. The SN wants tight IMPACTS bonding to be in service to it, and it wants to be able to control the release of that energy. In varying degrees, women are still mostly kept in their place around the globe. That is because their energy, like any IMPACTS energy, can be dangerous if not controlled. IMPACTS energy is new-beginnings energy, and an SN favors the status quo where it feels it has a tighter grip. But males are not as free as they would like to think either. From the moment of birth, they are pushed toward service to the SN and its philosophy.

In today's world, the SN in charge, no matter where it is located around the globe, is quick to use the incendiary term 'terrorist' if someone is vehemently expressing their dissatisfaction in some way, no matter how credible the grievances. SNs become islands unto themselves, unable or unwilling to see beyond their own selfish needs. We have to keep in mind that SNs are nucleated entities; therefore, they try to strengthen themselves and eliminate potential in adversaries. But these adversaries can quickly become friends, depending on what is needed or preferred at that point in time. The SN changes the rules as needed, and expects—actually demands—that the energy within its domain follow.

We have previously not understood how human civilization works because we have not thought of it as being composed of two parts, one part producing and the other accreting, just as is the case in the atom. Our problems are not usually between cultures and people at all. They are mostly generated by one SN against another, or between opposing blocks, as one or both try to attain more power and control, including resources. They enlist all the IMPACTS energy they can summon in their tasks.

Another reason we cannot see what is happening is because of what I mentioned earlier; we have been bred through natural selection not to see it. We think we are looking at the world through 'free-will' eyes. We are not. We are looking through eyes that have been naturally selected by the SN over the past 5,000 to 10,000 years. But nature knows the process at work. That is why she keeps creative-formative-productive energy—IMPACTS energy—on the periphery in her bag of tricks. She knows that if she does not, the SN will construct a homogeneity that will destroy CFPE and everything else.

The microcosm of what has been going on around the world is this: To a large extent, the hierarchal SN has gradually broken up the creative-formative-productive energy that was manifested originally in the San and San-shaman, and then in the IMPACTS.

The same happened with the development of the universe. First, there was the Big Bang of course, followed by billions of years of creative-formative-production. During this process nucleated galaxies were forming, each capturing creative-formative-productive energy in the form of stars and star-making potential. Now these galaxies are moving away from one another, though some will still intersect and merge. SNs are doing much the same on earth, each gathering as much IMPACTS 'star-making' energy as possible. Sometimes they too will merge, often violently, just as is the case with colliding galaxies.

The SN, though, does not want only IMPACTS and their innovative energy; it wants IMPACTS and workers. The original innovative energy of the San-shaman was based on the needs of people. The SN wants innovation based on its needs, which are structural. The SN just wants to grow—it is male energy. As it does, the bonds holding IMPACTS energy together can become stretched and tenuous. If activation energy levels are reached (2nd Law of Thermodynamics), the bonds can break and society can unravel. It has happened many times in history.

One primary element of today's world is not based on the San or San-shaman, and that is politics. That is why politics, like war, seems so unnatural, so adversarial and conflicted, so dark and divisive to many of us—it is not a San invention and therefore has few IMPACTS characteristics. Politics and war are mostly post-agricultural developments. The closer you get to the real power-control centers, the thinner the IMPACTS energy gets, generally speaking. But keep in mind that IMPACTS energy is powerful, and a little can still effect change. There appears to be significant IMPACTS energy within Barack Obama and around him, though there are intense power-control elements around him as well. But that is what we would expect. Obama is a politician too and politics is about power and control.

Interestingly, we find that many shamans in North America were more secretive than were San-shamans, and more feared. They were sometimes aligned with secular leaders also, placing themselves in the political arena. The leader could utilize the 'spiritual' powers of the shaman to increase his power, control, and wealth, and the shaman could obtain protection—theoretically. This is the model in operation by the SN around the world—'spiritual shamans and secular shamans' are enabling the SN to survive and prosper. Many modern countries also believe that they have God on their side as well.

In the Aztec empire that was based in present-day Mexico, thousands of people would be lined up to have their hearts cut out and offered to the sun god. Usually these were prisoners of war, another reason for the chiefs to conquer other lands. This kind of activity would most likely not have been possible without the help of shamans. After all, if they did not help the chief, they knew what awaited them. The pressures today to get in line are more subtle.

Connections

The San could see the connections among people everywhere and the need for those connections, and took it for granted. Why can't we see that the San and their genes were the ancestors of people around the globe?

We do not see it because the existing paradigm and the forces that benefit from that paradigm do not want us to see natural connectivity, nor do they want us to see cause and effect. A hierarchal world, or hierarchy of any kind, abhors a natural systemic connectedness unless the hierarchy installs that connectedness itself or can tap into the power of that connectedness and control it. A natural systemic cohesion can interfere with the flow of energy and power to the top, and a hierarchal structure wants no impediments to that flow. Anything is fine with the hierarchy as long as it enhances its power and control. Anything less than that is of little value or worse—an obstruction. If people today start seeing the natural connectedness of each other from long ago, that could spell problems for a power-control hierarchy.

One word best describes the attitudes, goals, and actions of the shaman and that word is transformation. And as we will see in the next chapter, the shaman was ready and willing to 'travel' far and wide to aid that transformation. Think about the valence electron—what is the result of its actions? It is transformation of the atom and the environment, resulting in more realization of potential.

It is also conservation of energy. Was life just the natural progression of bonding to conserve energy? Possibly. If so, everything worked pretty well until about 6,000-10,000 years ago when the capture of the IMPACTS by the SN began. With capture came the severance of powerful bonds and the redirection of energy, leading to dispersal and much destruction. But as we have seen, the process started long ago as evidenced by the herdsman ants-mealybugs relationship. With the development of the SN in the human sphere, nature had reached a point where the accretor could do some real damage to the earth and life. The human black hole had formed.

Summary

The first shaman (before the formation of the San) may have been the progenitor of homo sapiens and of the autistic spectrum as well.

The San-shaman and the San people in general exhibited an ego that was other-oriented, almost selfless. They were egalitarian and loving and strived for harmony with nature. The shaman was the leader of the tribe and the main problem-solver though there were often many shamans within the group, men and women but usually more men. The San tribe, however, was awash in female energy with very little male energy. The tribe was filled with innovation, caring, attention to others, cooperation, sharing—a very positive atmosphere for human survival and development.

Individual empowerment was encouraged; individuality with its expressions of self-importance was discouraged completely and totally. There is a big difference between individual empowerment during the time of the San when it served the survival of the tribe and what passes for individual empowerment today. Today, it is the glorification of the individual and the accumulation of possessions, which ultimately serves the SN structure and its leaders.

The San and the San-shaman formed the genetic blueprint for modern humans. Almost everything that exists in our modern culture, from business and trade, to art and music, to science, medicine, philosophy, and religion, originated with the San tribe, ultimately with the San-shaman. The foundation for any success we have had was laid a long time ago. We are the beneficiaries of their hard work and struggles, and of their positive outlook on life.

Footnotes

[1]*The Mind in the Cave*, p. 142

Chapter 5

The Shamanic Trance, Art, and More

Transformations were always needed: from sickness to health, from drought to rainfall, from poor hunting to plenty, from dissension to harmony, from disorganization to coherency, from imbalance to balance, from hopelessness to optimism, from unusable to usable, from marginal to the best it could be. In other words, the shaman was often consumed with the desire and need to optimize the potential of everything around him, and to discover the means to do so, whatever it took and wherever it led him. To facilitate these transformations, the shaman would have to transform himself from a person in ordinary reality to a seeker of solutions in the spirit world, where only he and other shamans could travel. The San believed in a tiered universe—a world below our material world and a world above. And you wondered where the concept of heaven and hell came from. It did not come from the San, but the model did.

How did the shaman enter into a trance? South African author and anthropologist David Lewis-Williams discusses the process in *The Mind in the Cave* and in *Inside the Neolithic Mind*. Repetitive sound and motion were tools. These could be expressed in music, chanting, singing, clapping, and drumming. Trances could also be induced through sensory deprivation, prolonged social isolation in dark places such as a cave (possibly with flickering lights), sleep deprivation, rapid breathing, searing pain such as a migraine, fatigue, focused concentration, hyperventilation, intense emotion, stress, food and water deprivation, even pressure on the eyeballs. The shaman would utilize those methods that worked best for him.

In some parts of the world, there is evidence that psychoactive plants were used by shamans to induce trance, but there is no evidence that they were utilized by San-shamans. The San people were the first; they were purists, and purists were needed to build the solid foundation for all of humankind.

According to Lewis-Williams, the stages of a trance are universal; the process works the same in all human brains, depending of course on physiological and psychological factors. The geometric forms seen first—dots, zigzags, grids, parallel lines, concentric circles or U-shaped lines, filigrees (meandering lines)—are actually patterns of neurons firing in the visual cortex. The normal state of the brain has been destabilized—the person in trance is seeing the brain's reaction.

Let's note that the journey begins with destabilization, or imbalance, or turbulence. Turbulence is seen within the mind of the shaman, and turbulence in the environment is generally the reason for the journey to the spirit world for solutions. Bonds have been broken or are in danger of being broken. In order to restore healthy bonding, the shaman has to first break his bonds with everyday reality before he can begin his quest.

This also has relevance to the autistic discussion we have been having. I have noted how innovative those with Asperger syndrome can be. There is a certain detachment from reality that allows for inventiveness. The theme is the same for both the shaman and Asperger—solutions reside in an area that is somewhat detached or very detached, or we could say is on the periphery. It is the exact same situation with the valence electron. Everything has to start somewhere and it appears that most of what we see around us started with the actions of the hydrogen atom's valence electron. It too was and is trying to initiate bonding, and it too is semi-detached from the structure, the atom.

In the second stage, Lewis-Williams notes that the brain tries to make sense of what is happening. The geometric forms morph into objects or animals that have emotional and cultural value to the person, or possibly represent a current concern. With the San-shaman in southern Africa, the geometric forms might take the shape of an eland or other animals in the environment, or perhaps the shape of a honeycomb since bees are a symbol of supernatural power for the San.

In the third stage, the shaman enters the Underworld through a vortex, or rotating tunnel. Objects become real. There he meets up with his animal helpers/guides such as the eland. The person in trance becomes a participant in and an inhabitant of the Otherworld, or spirit world, rather than a spectator.

Certain animals guarded the power of shamans and humans; a person would have his own animal guardian. It was believed that man and animal could combine into human-animal hybrids or actually trade places with each other—transform themselves into the other—at the spirit level. By doing so, a shaman could better understand how an animal behaved, which could facilitate the hunting of that animal. It was believed that the shaman could actually lead the hunted animal to the hunter, sometimes by singing. The southern African San believed that the potency of the eland could be acquired just by hunting the animal. This potency could then be used in a healing or trance dance.[1]

San Tribe and San-shamans

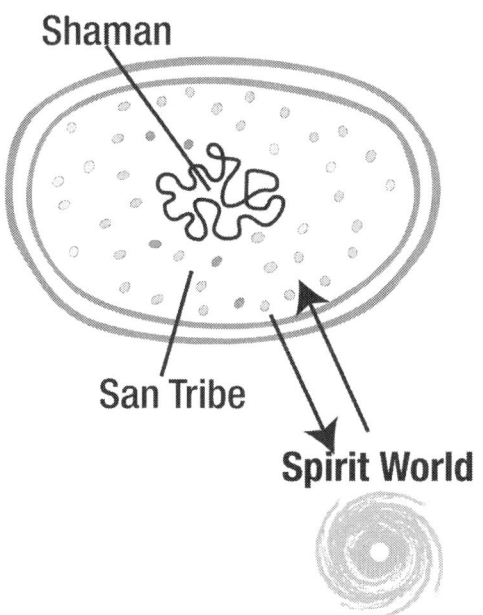

Shaman

San Tribe

Spirit World

Spirits, Animal Helpers,
Solutions

The San tribe and shamans accreted energy from the spirit world, animals, and the natural environment. The shamans traveled to the spirit world and returned with solutions.

The trance was also a journey of death and resurrection. The shaman had to suffer and 'die' first; then he could travel to the spirit world for solutions. Resurrected, he would return to earth with the answers. Christianity and the story of Jesus were built on the same theme. As mentioned earlier, Jesus was a 'resurrected San-shaman' as have been many other religious and social leaders. But that is no surprise because I believe that the San and San-shaman view of reality and the approach to that reality are now embedded in the genome and have been for the past 100,000 years or more. Humanity is invigorated every day around the world with thousands upon thousands of births of 'San and San-shamans', who are now the IMPACTS. They enter a powerful SN-guided world and their lives are sometimes torn between helping the structure or changing it. Mostly, however, their primary concern is helping people.

Now we can see why death in many religions and cultures meant that souls would travel to the spirit world. After all, that is what happened to the shaman when he died during trance. Keep in mind that just about everything we have in the modern world came from the San-shaman, including religion.

The Ju/'Hoansi (!Kung), a San tribe, describe the energy n/um produced in a trance as a burning liquid at the base of the spine, which then boils as it travels up the spine, exploding painfully out of the head. Sounds like the description of a volcano which also forms around conflicting forces and which provided early humans with many benefits.

The trance ritual was a social occasion as well. People laughed together, renewed bonds of friendship, and sang and danced. Not only did women protect the shaman from falling while in trance, they also directly influenced the strength of the healing energy, or n/um, by clapping, dancing, and singing. That brings to mind Sunday services in many African-American churches. The scene around the San campfire has just been transported. That statement is meant as nothing but a positive, to show that the same powerful forces are still at work. They are at work in other churches as well though the energy may not be as expressive. What are men and women in the choir doing in any church? They are augmenting the power of the sermon. That is why many of those who sing in church will be IMPACTS. It all goes back to the San.

The shaman knew that the answers that the tribe needed would not be found in ordinary reality. Feeling a responsibility to help meet the needs of others, he would have to build a bridge to the compassionate spirit world where his soul would travel for power and knowledge not found in everyday reality. He was the middleman, the medium, the facilitator, the transformation artist. He was the link between the known and the unknown. In his world, the shaman looked for connections, underlying meanings, patterns, threads of knowledge—anything that would aid him in his task. And it was an ongoing search, as the demands were continuous.

Everything had to be studied because the answers could be anywhere. The heavens and the motions of celestial bodies were studied; so too healing properties of plants and the meanings of dreams and visions. The quest was spiritual (in another realm), and it was also rational. The shaman was the first scientist, collecting empirical data and formulating theories, trying to discern the order that existed in the universe, much as I and millions of others are trying to do and millions before us have tried to do. If he could determine this order and figure out how the world worked as it did, then he could apply the newfound knowledge in ways that would aid his community and other groups nearby. That same attitude is seen in today's IMPACTS as they continually try to find real world applications for their ideas.

San-shaman Art

What I am trying to do is make a clear case that the early San tribe, with the shaman at the center as the primary problem-solver, was the foundational model for modern humans. As we will see, this shamanic model was then carried around the world when a group of the San made their exodus from Africa. With an embedded genetic blueprint and mechanisms for passing along cultural customs and knowledge, also the responsibility of the shamans, basic duplication of behavior, attitude, and philosophy was possible for tens of thousands of years. The shaman, like the valence electron, was the principal bonding agent for the tribe and also the conserver of energy through his ability to handle so many responsibilities.

163

The shaman's experiences in the spirit world often became part of everyday reality. At some point, what we call art became one of the shaman's primary means of delivering the power and knowledge from the spirit world. The paintings were in effect stored energy that could be harnessed by people at any time for generations to come. To the shaman it was not artistic but rather functional. To us, it was the derivation of art as we know it. The shaman, along with his functional art, was the bridge between the everyday world and the spirit world.

San-shamans over many thousands of years made southern Africa the largest art gallery in the world, replete with rock art and cave paintings. Not only are the paintings ubiquitous in southern Africa, they also exhibit a remarkable elegance and sensibility. There are about 15,000 known San rock art sites in the country of South Africa, with perhaps 50,000 in southern Africa. The oldest paintings, which were found in Namibia on rock slabs, are dated from about 27,000 years ago. The next oldest are from the Cave of Bees in Zimbabwe from about 10,500 years ago. But it is almost certain that San shamans were painting tens of thousands of years earlier.

San artists in southern Africa especially prized a pigment made from red ochre or hematite (iron oxide) that produced shiny, sparkling images. This pigment was found high in the Drakensberg Mountains, which would have required a lengthy hike to obtain. Men or women would heat the ochre and then grind it into a fine powder. Then it would be mixed with the blood of a freshly-killed eland, but only an eland. In that way, the potency of the eland could be captured within the paintings and then withdrawn as needed by placing hands on the paintings.[2]

I have absolutely no doubt that the inevitable development of pottery and ceramics, and later iron and steel, emanated from the San's early use of fire to heat red ochre for pigmentation. Somewhere along the line the discoveries were made by their descendants accidentally, which is the way most discoveries are made.

The journey to the Drakensberg Mountains to acquire a special pigment is an example of what we will see with the San and their descendants around the world—they will travel long distances or engage in Herculean efforts if a specific material is needed to complete the task, whatever it may be. This demonstrates a commitment to precision and to excellence, which the SN would later use to its benefit.

Lifting Stones

Let's look briefly at a few examples of this commitment to precision and excellence found around the world. At Stonehenge about 4,500 years ago, the builders in one phase of its development transported tremendous tonnage of bluestones from the Preseli Hills in Wales 240 miles over water and land to the Salisbury Plain in southern England. What was the purpose of Stonehenge and other such henges found in Europe, especially present-day Britain? They were probably healing centers and ritual centers, which included appreciation offered to ancestors and spirits. At the time of construction, Stonehenge was in the midst of the 'breadbasket' area. A more sedentary life gave these San-like descendants time to build impressive structures. The southern African San most likely used rock shelters for the same general ritualistic purposes.

At 9,000 feet in the Andes of Peru, a pre-Incan people constructed the city of Machu Picchu, carrying some stones weighing 200 tons up the mountain, and then placing one on top of another right on the edge of a precipice. We noted earlier that the stones fit together so precisely that even today a razor blade cannot be inserted into the mortar-less joints. How is that possible? Nobody has a clue.

Also in volcanic Peru, the Nazca Desert, which has had no appreciable rainfall for the last 10,000 years, is home to hundreds of lines hundreds of miles in length, portraying everything from geometric shapes to humans to birds, spiders, fish, and other animals. These lines, some of which extend into the Andes Mountains, were constructed by the Nazca culture from about 200 BCE to 700 CE by removing the top few inches of the reddish-brown iron oxide (ochre) coated pebbles that cover the desert floor. Many believe that some of the lines are related to astronomical or astrological concepts. As early humans spread out around the globe, they sought out environments that 'were in their blood'. That is why we see volcanoes and red ochre over and over.

Not only did the Nazca culture produce beautiful pottery, gold jewelry, and intricate weaving, but 1,000 human skulls have been found with holes that strongly suggest that brain surgery was performed! Why would brain surgery be necessary on such a large scale 2,000 years ago? I suggested earlier that 'psychiatric' conditions including Asperger's syndrome-autism may intersect with portions of the IMPACTS profile. That does not mean that all IMPACTS are afflicted; the vast majority are not. But there may be a correlation. What suffering were the Nazcans hoping to alleviate? Was it autism?

In the country of Bolivia located southeast of Peru, there is a place called Tiawanaco which was built at over 12,000 feet in the Andes. Pieces of volcanic rock, some weighing 150 tons, were transported from 200 miles away. Pieces of stone weighing 400 tons were also used, some of which were 'stapled' (bonded) together with some kind of molten metal. It appears that Tiawanaco may have been constructed as many as 10,000 years ago or more.

On Pohnpei, a tiny island that is part of Micronesia located in the Pacific about 1,000 miles north of New Guinea, is a city known by the natives as Nan Madol for 'City of the Gods'. It was constructed around 1,500 years ago on 92 man-made islands covering 11 square miles. About 250 million tons of basalt logs were used, sometimes stacked 40-50 ft high and weighing 50 tons each. Probably no more than 25,000 people lived in Nan Madol at its height.

Some natives say that the stones were levitated into place, while others say a pepper plant gave the workers extraordinary strength. Archaeologists believe that ropes made from hibiscus vines were used to pull the stones along on tree trunks before they were somehow raised into place.

How was such stone work possible in the areas just discussed and other areas around the world? As noted earlier, human beings developed in volcanic areas in the Great Rift Valley where there are extensive rock formations and limestone deposits. Plus, early modern humans attached a reverence to stone because they believed that spirits resided there. Stonework was probably 'in the blood' of modern humans for tens of thousands of years. But what these people had that mattered most was innovation, and innovation traveled wherever modern humans traveled. The artists were also the artisans, and those were shamans.

Paintings in Different Parts of the World

You will recall our prior discussion of the shaman's three-stage trance model. When anthropologist David Lewis-Williams compared that model with San art, he found all six entoptic signs in stage one (dots, zigzags, grids, parallel lines, concentric circles or U-shaped lines, and filigrees [meandering lines]) in their art. The evidence pointed to San-shamans as the artists.

Lewis-Williams once met an older woman whose father had been a shaman. She demonstrated how dancers seeking power turned to face the paintings on the wall of the rock shelter, and how some placed their hands on the paintings of eland to gain power. The paintings were repositories of energy available to those in need.

The eland is the most frequently depicted animal in San art and the most revered. Why would that be the case? The eland held a special position in San life and opened the power of the spirit world for the shaman. The eland is the largest antelope in the world, its weight sometimes exceeding one ton. Its meat is very tasty and has the most fat of any animal in southern Africa, very important to a foraging society. Beautifully athletic animals, they are sometimes seen jumping over each other. But they are not blessed with great intelligence, or so it has been said. All of these factors combined to make them the perfect animal of prey for the San—and a lifesaver for them. Today we might say they were heaven-sent. To the San, they were spirit-world-sent. The San treated the eland and all their prey with great reverence and appreciation.

Other indications of the trance in San art: depictions of nasal bleeding, dancers bending forward with their hands thrown backward, the use of sticks in each hand to support the shaman's body, mythical and 'hybrid' animals, and the partial or whole transformation of the shaman into the therianthrope—part man, part animal. The rainmakers were drawn: hippos, elephants, and giraffes. There were also underwater images seen during the trance along with drawings of animals suspended in air as if flying through the spirit world.

Rock shelters almost certainly became places of veneration where bands of San gathered for so-called aggregations, sharing rituals and initiations, celebrating the abundance of food, telling stories of tribal history and mythology, and discussing life and exchanging ideas for dealing with it. In widely separated rock shelters, the paintings would be almost exactly the same, revealing a close collaboration among the shaman artists along with similar skills and materials utilized, and of course similar genealogy. An overall stasis was in place.

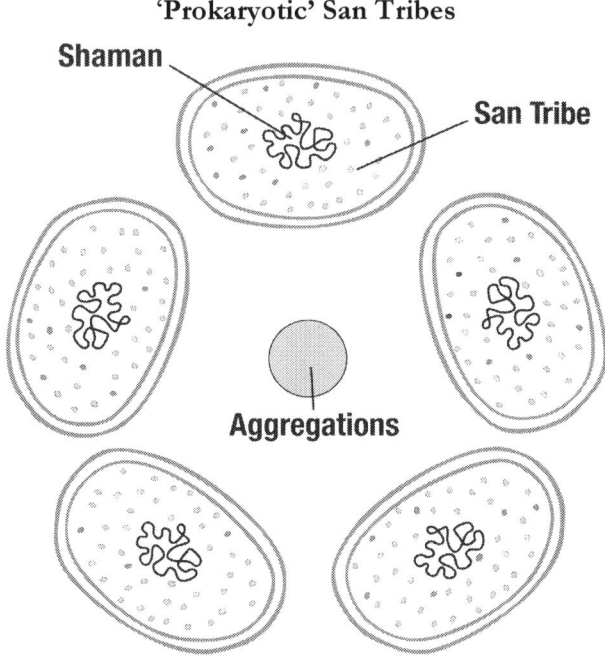

Aggregations Provided a Setting for the Renewal of Bonds, the Exchange of Ideas, and the Continuation of Tribal Customs

The interactions of the various bands of San followed the 'prokaryotic' model. It was not a centralized network but more like a network of independent nodes. This was the blueprint for human proliferation and emerging civilizations around the world—nodes built around San-like descendants, or the IMPACTS. As SNs developed, they would attempt to draw these nodes under their 'structural-nucleus' umbrella. The same is still occurring today.

Formation of 'Eukaryotic' Society and the SN

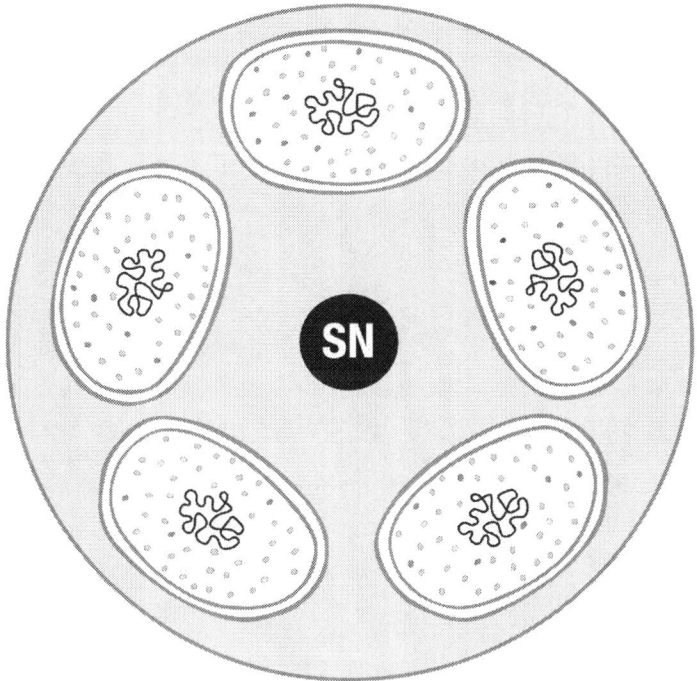

The SN Became the Nucleus of the 'Human Cell' Using the IMPACTS and IMPACTS Nodes as its Mitochondria

The San paintings were never considered to be complete as images were painted one on top of the other over and over again. Thousands of years later, San-like descendants would take the same attitude towards the building of their houses—they would build one house on top of another, sometimes several times.

It appears that cave art found in European caves, art known as Upper Paleolithic (40,000 years ago to 10,000 years ago), is also shamanic in origin. The three-stage model fit the Upper Paleolithic art just as it did San art of southern Africa. In the Upper Paleolithic in Europe, it appears that the horse and bison were the sources of potency, whereas the eland had been for the San. But the Upper Paleolithic art, like the San art, also included images seen in the third-stage spirit world: therianthropes (part human-part animal), 'monsters', and realistic animals.

In *The Mind in the Cave*, Lewis-Williams relates how he found many of the same shamanistic images and motifs in the rock art of the San tribe !Kung of southern Africa, the rock art of Native Americans, and the cave art of Ice Age Europe. In addition, the !Kung and other San tribes see stone as a porous membrane separating the spirit world from the everyday world. By painting on rock and stone, shaman artists were communicating with the spirit world, and the spirit world was also communicating with them. Again, the shaman-artist was acting as the intermediary and facilitator of communication between the two worlds. He identified the spirits as the real artists, working through him.

On Nigel Spivey's DVD, *How Art Made the World*, he and David Lewis-Williams discuss how the European cave paintings from thousands of years ago have a remarkable resemblance to paintings in the Drakensberg Mountains of South Africa, painted just 200 years ago by the San. That is not a coincidence.

Some of the oldest paintings on the planet have been discovered around Oenpelli, a small settlement in Arnhem Land in northern Australia, dated at 40,000-50,000 years ago. These are the first known art galleries in the world. Spivey notes in his book, *How Art Made the World*, that the quantity of the San paintings in the Drakensberg Mountains of South Africa is about the same as that in this particular area of Australia, another clue to the genealogy of the painters.

The Englishman Baldwin Spencer studied the Aborigines around Oenpelli, whom he called the Kakadu, in the early twentieth century, and found them to be obsessed with painting. The same is true today. They paint the same images over and over, the same ones found on rocks and hills from thousands of years ago. They appear to be obsessive-compulsive, a trait we often see among IMPACTS and autistics as well.

The paintings are part of a story; they are understood only within that context. The symbols trigger memories of the story. The paintings and the storytelling are accompanied by music and dance. This has been the principal manner of passing down stories through the generations.

Art was a means of communication for early modern humans just as it is today. Bonding was maintained with the spirit world, memories were kept intact, culture and mythology were transferred to future generations, and sheer enjoyment was experienced. As noted previously, art and the shaman were the bridge to another reality, a reality that offered potentialities that could be meshed with everyday reality. They were the intermediaries between the real and the possible. It is the same in the world today with the IMPACTS.

Out of Africa

All evidence points to the San as the out-of-Africa group. I cannot imagine who else it could possibly be. The fact that it is not accepted knowledge shows the power of the SN to sculpt our view of reality. Just imagine what other parts of reality are being sculpted. What we see everyday is mostly a product of SN-engineering—a top-down way of looking at everything which is directly opposite of the way everything develops—from the bottom up.

The Australian Aborigines are direct descendants of the San who left Africa 60,000 to 80,000 years ago or so. That is why their lifestyle and paintings are so similar. If you have a group of 1,000 golden retrievers which leave Africa and travel to Australia, and they mate only with one another because there are no other dogs anywhere along the way, then when they arrive in Australia it will still be a pack of nothing but golden retrievers. And they will still be behaving just as they did when they left Africa.

Leaving Africa was not a Noah's Ark event. Two from every tribe were not gathered together. It was a group from one tribe, and by all indications that was the San.

This is not complicated. It is as clear as the nose on our face. If we showed it to a group of third graders and didn't try to influence their deductive processes, they would probably see it clearly. Adults don't see it because we have been taught not to see it, and because no one has brought it to our attention. The paradigm warps our thinking just as Einstein said that mass and energy distort space-time. We have deferred to the paradigm to define the 'truth' for us, and the SN has taken full advantage. It is astounding how easily the SN succeeds in getting us not to see the real picture.

It is interesting that the Aboriginal way of combining storytelling, painting, and music was not practiced during the ancient classical civilizations. As Spivey notes in *How Art Made the*

171

World, it was not until the religions understood the power of music combined with symbols that it was resurrected. That is amazing when you think about it, that it could have been buried for so long. It has also not been that long ago that movie production companies learned what religions had learned—add music to the mix and everything changes.

As civilization developed, the SN made a concerted effort to rid society of any vestiges of shamanic behavior. It wanted the skills of the shaman but not the behavior. As SNs went about their brutal business of natural selection, they destroyed the dynamic that housed the stories and pictures and music, which was the shamanic culture. They broke the 'chemical' bonds that held it all together. But that is no surprise; that is what hierarchies do. They chop it up and take what they want.

It takes a long time and immense struggle for the natural configuration to finally mold itself back into something resembling its former self, if it is possible. As I have mentioned previously, human society still has not returned to the days of the San when men and women treated each other equally, and when justice and doing the right thing was the preferred way of life. We may never be able to assemble that kind of society again. The 'pool' has been contaminated by SN power-control elements and there is no turning back. Even in the material world, we are reinventing much that had been developed thousands of years ago but was then lost as the ascendant SN put a stranglehold on cohesion and innovation. That can be recovered because it has value for the SN; justice and doing the right thing do not.

What the San-shaman Gave Us

The shaman had many roles, but all of them were based on wellness for individuals and the community. That necessitated being a scientist first, carefully observing and studying life and nature in pursuit of helping aids. Plus, he/she was a singer, poet, artist, prophet of weather and hunting success, spiritual leader, diviner, sage, mystic, seer, magician, student of human behavior, and storyteller. The shaman was also the guardian-keeper of the tribe's traditions, practices, knowledge, treasures, and mythology, and the custodian of the calendar—the tribal time-keeper. Plus, ensuring that the souls of the dead traveled to their proper places in the spirit world was a shamanic responsibility as well. Why did he have so many important roles? Because he was trusted, and capable, and needed. This is why IMPACTS are attracted like a magnet to positions of responsibility; they get it from the San-shaman.

**San and San-shaman Genes Provided a
Strong Foundation for Modern Society**

The human female had found a partner who could aid in the survival of the human race, and that was a male with shamanic attributes. This opened up new possibilities for humanity. But just like the energetic, mobile electron, eventually this bundle of androgynous IMPACTS energy would be captured.

Shamanic tradition is without dogma, and without conventional wisdom. It is an ongoing, open search for better ways of interpreting and explaining reality, and improving it. This is the attitude on which the San-shaman profile was based and the attitude that forms the foundation of the IMPACTS profile.

What was the real ingredient in the shaman's repertoire that enabled healing? Was it knowledge and power from the spirit world, or were his actions actually a placebo that worked because of the closeness between the patient, the healer, and the remainder of the tribe? Surely the confidence, respect, and trust that existed among tribal members were important elements. But perhaps the most important was the deep transcending love that the shaman possessed for humanity and the tribe, the love that was the source of the motivation that drove him to find answers for the sufferings of others.

The shaman was willing to travel to other worlds for power and knowledge at a real risk to himself; some shamans actually 'lost' their mind on these journeys. He was searching for a creative healing force, and because he had love, he himself became part of that healing force. I believe that love is the antithesis of the 2nd Law of Thermodynamics—it is the strongest bond against dispersal of human beings. I think that is why we suffer so when we are separated from our loved ones—the bonds against dispersal have been broken or stretched thin.

There was a lot of love within the San tribe—for each other, for family, for the earth, for animals, for life, for existence. The San saw themselves as part of the uninterrupted chain of life. It all sounds very Jesus-like, doesn't it? But that is because all of the questions about life and existence and the afterlife originated with the San-shaman, or perhaps even earlier shamans. Most of the aspects of Christianity and every other religion can be traced directly back to the San-shaman.

The shaman has been portrayed in many different ways, very few of which have been flattering. But eccentricity is part of the peripheral nature of change. Forces within homogeneity do not generally change homogeneity. How could they?

The San tribe and the San-shaman, along with the trance, gave modern humanity a foundation for survivability. The San-shaman was the DNA of the modern species in that he was the repository of information that was needed to keep the species going. He was a duplicative force. The shaman was the inventor of the new and the guardian of the old. It is the same with the IMPACTS today.

The trance and related activities were extremely valuable for the cohesion, and therefore survival, of the tribe. It also gave us representational art and a spirit world that became an integral part of religion. But most importantly for our species, the trance embedded the search for answers in the genome of modern humans through the creation of the IMPACTS profile. But regretfully, the SN does not take full advantage of all of the gifts offered by the IMPACTS profile. It only takes what it wants and throws the rest to the periphery—or worse.

Great Leap Forward?

Anthropologists have noted that about 40,000 years ago there was a technological and cultural explosion in Europe. David Lewis-Williams in *The Mind in the Cave* says there was more diversity of raw materials, new tool types, regional tool styles, more sophisticated hunting strategies, organized settlement patterns, and specialized trade. There was also extensive body decoration, more elaborate burials, and portable and parietal artwork.[3] The area had some of the markings of today's world with apparent social tensions and social conflicts.

Here we probably see the early beginnings of an SN or hierarchal structure. The Ice Age conditions at the time would have limited mobility and produced the organized settlement patterns that Lewis-Williams mentioned. If IMPACTS are clustered in significant numbers and the settlements are fairly permanent, male SN structural energy has a good chance of developing and 'managing' the production of the IMPACTS. Think back to the mealybugs and herdsman ants we discussed in Chapter 1. An energy source existed in the environment and the ants discovered it and utilized it. That is the way it works throughout the universe—energy is produced and energy is accreted.

It appears that the modern behavior displayed in the Upper Paleolithic in western Europe was actually an extension of behavior that started much earlier—75,000 years ago or more—in Africa. Modern humans are believed to have left Africa 60,000 to 80,000 years ago. To undertake such an incredible journey and then to survive and prosper as modern man has done would require a powerful and complete arsenal of life-sustaining aids, or modern behavior. Personally, I believe that modern behavior began with the 'first people', the San, probably about 120,000 to 130,000 years ago.

In Blombos Cave on the southern coast of present-day South Africa, beautiful pearl-like beads dated from 75,000 years ago have been found. These beads, evidently used as personal adornment or gifts, were meticulously crafted from the shells of small snails belonging to the species Nassarius kraussianus. Similar beads found in present-day Israel and Algeria made from the species Nassarius gibbosulus appear to be about 100,000 years old. Other beads constructed from ostrich eggshells have been found by Stanley Ambrose in Kenya dating from 43,000 years ago. The same San group we discussed earlier, the !Kung of Botswana, are still making and exchanging beads as gifts that are strikingly similar to those mentioned above.

Also in Blombos cave, dated from the same time period, about 40 bone tools and hundreds of aerodynamically-designed bifacial hunting points were found. The carefully-polished endpoints entered the skin of the animal much easier because of lessened resistance. This same aerodynamic principle was utilized later in other parts of the world in the construction of the non-returning boomerang (hunting stick) and the returning boomerang. Thousands of years later the same principle contributed to the invention of the airplane.

Hearths for cooking were also discovered in the Blombos cave. Red ochre was found engraved with geometric designs, formed after the ochre had been precisely scraped and ground to prepare a nice flat surface for the engravers. These engravings are uncannily similar to some found recently in 150-year-old art produced by the San of southern Africa. Red ochre is also used to paint the human body decoratively. Thousands of pieces, some in crayon form, were found in the cave.

The oldest fossilized human footprints were found in the same geographical area as Blombos cave, at the Klasies River cave sites, dated at over 100,000 years ago. Also found were indications that fishermen had used boats many tens of thousands of years ago.

At three sites in Katanda, the Congo, harpoons carved from bone were found that dated from 65,000 years ago. The quality, exquisite design, and construction material were very similar to harpoons found in Europe dated 25,000 years ago. Everything about the harpoons suggested that what we might call the artistic presentation was just as important as the function. The work of the San indicates a similar attitude.

The harpoons were found with the remains of large catfish, which suggested that the fishermen were planning their catches during the spawning season. Many anthropologists had believed that this behavior came much later.

Finds in Tanzania in the Serengeti National Park are similar to those at Blombos Cave located hundreds of miles away—ochre pencils, bone artifacts, fish bones, mammal bones, and ostrich egg shell beads.

Elegant rock paintings from 5,000 years ago on the Tassili plateau in Algeria and the Gilf Kebir plateau in Egypt are very similar in color and style, again, to recent renderings of the San of southern Africa. The Algerian and Egyptian paintings depict a green Sahara with flowing rivers and cattle herders, a far cry from today's inhospitable environment. The paintings should be another indication of the proliferation of the San and a clue to the foundation of the Egyptian civilization.

In rock art around the world—the southwestern U.S., the Middle East, Brazil, Australia, central India, Sahara, southern Africa—the paintings or engravings all combine three basic elements: geometric figures or abstract designs (stage one trance), animals (stages two and three), and human figures. The animal drawings are the most beautifully naturalistic and precise whereas human representations are exactly the opposite. Interestingly, landscape renderings including plants, fruits, and flowers are mostly absent, but then they are not generally seen in shamanic trances. The art around the world appears to have been generally shamanistic in nature, depicting what was seen in journeys to the spirit world.

The shaman would not have 'seen' humans by and large in the spirit world; it was home to spirit guides and helpers, and those were animals. Of course human beings were important. After all, their health and well-being were the reasons for his work. But in his trips to the world beyond, animals were the potent force and therefore the object of his paintings. He painted what he saw.

Another aspect that must be stressed—the techniques, colors, philosophy, and tools utilized by prehistoric artists were remarkably similar around the world. To what do we ascribe these similarities? Why would art around the world be so uniform during so-called prehistoric times? Why would everyone be using the same colorants, one of the primary ones being ochre, the same techniques, and the same three basic elements of geometric symbols, animals, and humans? If we look at the image-making as the production of the same genealogy and culture, that of the San tribe and San-shaman, it starts to make perfect sense.

In the next chapter, we will talk about the exodus of modern humans from Africa and the gradual development of human society from the mobile hunter-gatherer to the more sedentary lifestyle enabled by agriculture.

Summary

The San-shaman was of two minds; he was part of the everyday life of the tribe, but he also traveled to the Otherworld, the spirit world, where animal spirits aided him in his quest for answers to tribal concerns. When he returned to everyday reality, he brought those answers from the periphery.

An important element here is travel. IMPACTS travel in order to find answers, just as the San-shaman did. We will see IMPACTS replicating practically all of the San and San-shaman's behavior. IMPACTS possess the same basic personality profile and worldview of the early San tribe, including the problem-solving aspects. But some IMPACTS, those closer to the San-shaman profile, are better problem-solvers than others.

To the San, art was not art the way we think of it—it had a function, just as hunting did. It aided communication between the spirit world and people, united generations around themes of life, aided in keeping the tribe close-knit, and provided a source of great pleasure. The San-shaman and art served many of the same purposes because they were so tightly intertwined. One could not exist without the other.

The most important word to describe the mentality of the San-shaman is transformation. His goal was to transform sickness to wellness, whether it related to individuals, to the group, or to the environment in general. Some IMPACTS will be more like the San-shaman and try to initiate transformation while other IMPACTS will be more like the San tribe members and help carry out the transformation, or deliver the results.

Footnotes

[1] *Inside the Neolithic Mind*, p. 116.

[2] *The Mind in the Cave*, p. 159.

[3] *The Mind in the Cave, Chapter 3.*

Chapter 6

Out of Africa to Agriculture

We have seen that different San tribes living in semi-isolation from each other can retain the same basic philosophy and behavior for 100,000 years. This demonstrates how deeply ingrained are the San personality traits and characteristics. You might say, "Of course they would remain the same. Same environment, same genes. There is nothing that would precipitate a change."

So if these same people, the San, left Africa and maintained their same shaman-centered way of life and their same customs, and they sought out the same kinds of physical environments in which they had developed over tens of thousands of years, why would it be any different if they spread out around the world? They would adapt to changing environments—skin color would change, some skills would be emphasized over others, new talents would emerge in different locales. But otherwise the basic core of the San would remain the same. The emerging human species had been built around and upon the specific roles, skills, and genes of the San-shaman. That was the developing structure. The foundation had been laid and that would not change. It remains the same today.

The San were exchanging genes with each other just as they had always done, except now they were spreading out around the world. They were the only group; they were not encountering others along the way. There were no others.

That is why everyone around the world behaved very similarly for tens of thousands of years. They were all San descendants, and they were all continuing to live the same basic lifestyle, based on the shaman. If you had one wolf species, and it had developed over tens of thousands of years in Canada, and then traveled to Alaska, and further on to Siberia, there is no reason to think that the species would change its 'pack' behavior in any way, if it were working. And of course the wolves would be seeking out environments that they knew.

"But we are not wolves," you might say. No, we are not, but we are not as different as you might think. Today's human society is markedly similar to the structure of a wolf pack. But when the San left Africa, there were two big differences between them and a wolf pack—there were no alphas and no omegas. In today's world, the post-San era, there is a plethora of each.

A wolf pack and contemporary society are nucleated, hierarchal structures. One of the reasons for a hierarchal structure is that it prevents in-fighting, which would deplete the structure of badly needed energy. As long as everyone accepts their role(s), energy can be conserved though change in the hierarchal structure can be difficult and slow.

With the San's egalitarian circular set-up, greater flexibility was maintained. Challenges could be approached with the full energy of the group, if that was what was needed. In a hierarchal arrangement, the top position sets the agenda, and input can often be limited.

Keep in mind as we go forward that we are dealing with very small population numbers for a long period of time, all of humanity at many intervals totaling maybe one-fifth, or less, of the average attendance at a college football game. The population may have declined on occasion to less than 5,000 people due to natural disasters such as the eruption of the Toba volcano in Indonesia about 70,000 years ago. It was the largest volcanic eruption of the past two million years. Who would have been best equipped to survive these types of catastrophes? The obvious answer is those with the most survival skills, and that appears to be the San, the first modern humans.

You will recall that homo ergaster used the same tool for possibly one million years. Nothing ever punctured the equilibrium of its existence. Yes, modern humans were different in many ways from earlier species, particularly with the shaman and innovation, but nature usually develops a stasis, and that stasis hangs on until something in the environment, or possibly a mutation, forces a change. The San today in Africa are still behaving very much as their ancestors did 100,000 years ago, or more. But stasis had set in long before the journey out-of-Africa began. How long could that stasis remain intact as the San spread out around the globe? What might change it? If it changed, what would be the new configuration?

Some 60,000 to 80,000 years ago, all indications are that a San group of around 1,000 individuals or less left the coast of present-day Ethiopia and crossed the southern end of the Red Sea, reaching present-day Yemen on the first leg of an epic journey that continues today. Why is it still continuing? Because human beings, especially IMPACTS, are constantly searching for greener pastures, and that includes new discoveries in any and all areas of human endeavor. As the world has become smaller, the journey is now more mental than physical, just as it was with the San-shaman.

It was only 10 miles or so between Africa and the Arabian Peninsula at the closest point, so most could possibly have waded across since sea levels were much lower at the time. Some boats may have been used. The San had probably already been in existence for 50,000 years or longer, and were most likely the group of modern humans that had fanned out across Africa about 100,000 years ago. There is evidence that a group also traveled to the Near East about that time, and again this was probably a San tribe. So the San were not afraid of new experiences and new challenges. In fact, they seemed to relish them. You could say that they had a 'new experiences' gene, which meant they also had a manageable fear of the unknown along with ongoing inquisitiveness.

Duplication

The group that left Africa would be attracted to areas similar to those where they had developed, where they had learned to survive. Their natural environment was of course a major part of who they were, and they would try to replicate that environment in their travels, just as any of us would do if we were setting out on a journey of the unknown. Their civilization had developed in a heavily volcanic area with fertile soil, caves, obsidian that could be used for tools, red ochre, limestone, natural springs, lakes and rivers, the sea, and plentiful game, and they would seek out the same general environment.

Again, if we understand the San and shaman community structure and roles, we will understand the development of human civilization. Everything was built on the foundation of the San just as everything in the known universe is built on hydrogen. Let's keep it simple because it is simple.

You cannot have a structure of any consequence without a creative-formative-productive agent or energy, and for modern humans that was the San tribe, especially the San-shaman. It was the same for the San tribe itself—the creative-formative-productive agent was the shaman. It is like a house—you cannot build the body of the house until the foundation is solidly in place. Without a solid foundation, there is no chance for a reliable, lasting structure. The main reason today's structure of human civilization is in such trouble is because the SN has been able to do what all nuclei attempt to do—it has 'eliminated much of the potential' of the foundational San energy (IMPACTS energy) by dividing it up, trying to control it in measured ways. Remember, IMPACTS energy is the fuel for civilization just as gasoline is fuel for a car, and gasoline gets utilized in very controlled increments. The SN can't be nearly as efficient but its goals are the same.

It is important to see the departing group of 1,000 or so for what they were, essentially an extended family. There were probably only a few thousand people living at that time, and 1,000 of those were leaving for other lands.

Other human species had left Africa but had ultimately not been able to colonize the world as the San were about to do. The critical difference was the shaman and the innovation that he/she possessed and passed down, genetically and culturally, from generation to generation. The model which had helped the San survive for thousands of years—close-knit, cooperative, sharing, and shaman-centered—would be put to the ultimate test. But I am certain they began their journey with the utmost confidence of success and survivability. Why wouldn't they? The model in place had passed the test of time along with almost every conceivable challenge.

The Journey of the San

Some of the travelers followed the coastlines of Arabia, India, Southeast Asia and Indonesia, and on to Australia. That journey of 9,000 miles or so may have required 10,000-15,000 years to complete. Evidence suggests that humans arrived in Australia perhaps 50,000-60,000 years ago. There again, they were following what they knew—the coastline. They could survive there: a temperate climate, food from the sea, and game that could be hunted inland. We can assume that sometimes a subgroup would split off and head inland upon approaching a delta region—but only with a shaman. Survival demanded that the well-tested structure of the community be maintained. Keep in mind that the San were branching off from the San. They had the same genes, same ancestry, same customs, same knowledge, same philosophy, and same basic behavior.

IMPACTS energy usually comes from the periphery in order to effect change. But it is also manifested by those who go to the periphery—it is two sides of the same coin. The emigrants are the immigrants. The periphery attracts the IMPACTS just as it did the San. Like the San, IMPACTS really just want to be left alone so they can explore, discover, share, innovate, create-form-produce, and deliver the solutions to problems. As the San and their descendants spread around the world, they sought privacy and peacefulness such as the San had enjoyed in their ancestral lands. But as time went on, the SN-forces would arise and make that goal very difficult to sustain.

Modern Human I

Keep in mind as we go forward that when we talk about the characteristics of the San and San-shaman, we are essentially talking about the characteristics of today's IMPACTS. IMPACTS are not homogeneous just as the San tribe was not homogeneous. Some are more like the problem-solving shaman while others are more akin to helpers of the shaman who aid in the delivery of critical solutions. And we have also seen that there are SN-IMPACTS closer to the mainstream and P-IMPACTS more comfortable on the periphery.

The San tribe with the shaman at the center and their descendants who adhered to the same basic attitude and philosophy toward life and people are what I call Modern Human I. Today's IMPACTS are foundationally Modern Human I, but like mitochondria and the nucleus of a cell, a large percentage are now tightly bound with the SN and Modern Human II, which we will get to in a moment.

The San spread out around the world doing what the San had always done—surviving through cooperation and resourcefulness, using their artistic abilities for functional art and tool-making, and living in a close relationship with nature. During ice ages, travelers would often live at the edge of ice sheets where conditions might resemble those today on the Alaskan tundra—cold but full of game. Others would settle in refuges such as those in northeast Spain and southwest France, where many caves have been discovered with phenomenal paintings. There is no mystery about why these paintings are so similar to others found around the world—the artists are carrying San-shaman genes.

Some groups remained south of the ice sheets in more temperate areas, such as the Mediterranean and the Aegean. These regions provided much of what people in that period needed for survival: rugged coastlines that provided seclusion, quick access to the sea 'highway', a reliable source of food, and a volcanic environment that was very similar to modern humans' beginnings in Africa. Around volcanoes, they could find and utilize obsidian for tools, enjoy natural springs, and know that the fertile soil would spawn rich plant life and hence attract plenty of game.

Interestingly, archaeological sites in South Africa dating from 60,000 years ago to 30,000 years ago show a diminishment of 'modern' behavior. There is evidence that the population declined significantly about that time, which also coincides roughly with the out-of-Africa movement. Remember, populations were very small—thousands, not millions. It would not take an exodus of many before a decline in modernity would be clearly discernible. Many of those who left for new adventures would have been the 'best and brightest' just as is the case today.

European Art and Artifacts

During the last Ice Age over 20,000 years ago, 10,000 years before agricultural development began, textiles and baskets were being made at Pavlov and Dolni Vestonice in what is now Moravia, the Czech Republic. Evidence of ceramic technology was also uncovered at Dolni Vestonice from the same time period. Beads and other personal decorations were also commonplace, even in death. One male skeleton was found in Russia with 3,000 ornamental beads around it, along with 25 mammoth ivory bracelets on its arms. Along with the older man's skeleton were those of a young boy and girl, each with about 5,000 beads around them. If each bead took forty-five minutes to make, it would have required over 7,000 hours of work.

Would the community have done this for every member, or is this evidence of a hierarchal structure? Or is it a shaman and possibly two of his children, or two other children from the community? It is quite possible that they died in an accident.

Beautifully crafted 'Venus' figurines have also been found in the area dated from 20,000 to 29,000 years ago, with most being around 4 inches in length. Made from bone, ivory, soft stone, and pottery, they were unmatched in their craftsmanship for thousands of years afterward. Ice Age Europeans were apparently making ceramic figurines over 12,000 years before pottery bowls and cups were crafted in Japan.

Probably the most famous figurine, that of Venus of Willendorf (Austria), was made of limestone coated with red ochre. (There is the recurring red ochre we see so often.) Actually, hundreds of similar figurines have been discovered in an arc from the southern British Isles all the way to Siberia, an arc which probably followed the glacier's lush edge and its abundant game. Recently, a similar ivory figurine was unearthed in a German cave and dated at 35,000 years, making it the oldest one found so far. The crafted figurines look basically the same because the human groups are still very San-like.

Venus of Willendorf – Photo by Matthias Kabel

Think back to the San tribes and how they lived. They were extremely independent and resourceful but interdependent as well. They traded with other San tribes and also set up hxaro exchanges which promoted goodwill that might be needed when times were more challenging. When conditions were really tough, a band might split up and the members join up with other bands. Most likely a similar template was still being used during the Ice Ages in Europe. It was basically San tribes from Africa relocated to Ice Age Europe. The communities in Europe were probably connected 'nodes' just as the San were in Africa, and connected in the same way—through shamans.

Living in our present world of extreme individuality entangled with a nucleated structure, it is hard for us to look back 25,000 years and visualize what was most likely a very clone-like structure of humanity, much as the San still are today. But there is no reason to think it was otherwise and many clues that it was exactly that. In our 2nd Law of Thermodynamics model, the bonds that had been formed by the San tens of thousands of years earlier were still intact and were keeping dispersal forces at bay. What dispersal forces? Dispersal forces exist everywhere in the universe. Bonds are the only things that keep everything from coming apart. The special human bonding which had originated with the San was the lifeline during extensive human migration. Today IMPACTS bonding is the same and performs the same job of holding humanity together even as dispersal forces try to rip it apart.

The figurines discovered across Eurasia depict the steatopygia condition prevalent until recently among the San of southern Africa. Similar 'Goddess' figurines were also found much later in the 'Old Europe' Balkan culture, and in Catal Huyuk in Anatolia, or present-day Turkey. Did the steatopygia condition follow the San and their descendants around the world, even in cooler environs? I do not know, but if it did it certainly explains a lot. If it did not, the descendants of the San people who had left Africa would certainly have had knowledge of steatopygia by passing down stories generation after generation. The San revered and worshiped their ancestors just as did those who settled around the world. Also, San men cherished San women.

Some anthropologists have expressed a belief that the Venus figurines symbolize fertility, and certainly that was a part of it. Many of the figurines show signs of being passed around. In *The Mind in the Cave*, Lewis-Williams noted that the Upper Paleolithic carvers who used animal parts such as ivory or bone believed that their figurines possessed the same powers as the animal out of which they were carved. Shamanic societies around the world also had the same attitude towards stone; the potency of the stone was released as it was being crafted. I would suggest the same kind of thinking applied to the Venus figurines: the potency of the material was released as it was being crafted, and was also stored within the finished product.

Again, we need to recognize that behavior such as this around the world is emanating from the original San genealogy and culture. So we are going to see the same basic behavior everywhere we look with only regional differences.

Arms are not present on the figurines. Among the San with steatopygia, the arms were very thin, and it would be easy to see why they would be deemphasized. The emphasis was on the parts of the body that had saved the tribe and the species—the midsection, including the breasts.

It is suggested by some that the Venus of Willendorf figurine is 'wearing' a basketwork hat. I do not think so. The San have a peppercorn hair structure that somewhat mimics the look on the figurine. That 'hat' is probably the peppercorn hair of the San ancestors of the Ice Age Europeans. Why was there no face on the figurine, just a head with the peppercorn look? Possibly because the San and their descendants focused their attention on the group and not the individual. There was no concept of individuality as we know it today among the San. You will recall from our earlier discussion of the San that even when hunting, the accomplishments of the group were emphasized over those of the individual.

In Lascaux Cave in France, a small seashell stained with the oft-used red ochre was found that had been perforated to be used as a pendant. The seashell had been collected from the Atlantic coast 120 miles away.[1] We have been assuming that the seashell beads we have seen before were mainly for personal ornamentation. To the wearers, they may also have possessed potency from the sea much as figurines did from ivory, bone, or stone. Also, note the distance traveled for the shells. Here again we are seeing San-like descendants travel great distances for the specific material needed. The whole process, the act of travel to collect the shells and then the creative-formative-production of the beads, reveals the precise nature we alluded to earlier.

Primitive flutes made from bone have also been found in western Europe from the Upper Paleolithic era.[2] Recently, a 35,000 year old bird flute was unearthed in a German cave. That is why you see many IMPACTS today playing classic instruments such as the flute, guitar, harp, and the piano, which is really an embellished harp. IMPACTS are playing instruments that were popular with Modern Human I.

Humming sound-makers called 'bull-roarers' have also been found in Europe. Attached to a cord, these were flat pieces of wood, antler, or bone swung round and round. One 'bull-roarer' decorated with geometric designs and stained with red ochre (there it is again) was found in the Dordogne region of southwestern France. The same instrument has been used by the southern African San to mimic the sound of swarming bees, which the San believe possess supernatural potency available to them.[3]

Can there be any further questions about who the out-of-Africa group was and from whom we are descended? I would think the evidence is overwhelming.

Cave Art in Europe, Mostly 18,000 to 10,000 BCE

By far, most of the cave art has been found in France and Spain, and the same elements are present there as are seen in San art in southern Africa.

- Geometric figures and designs
- Large numbers of animals in various activities
- Human beings, though rare and often as stick figures, or as part of a human-animal hybrid (therianthropes)
- Human hands

In European caves, soft clay on rock surfaces was the canvas for the earliest images, which were finger drawings. Engraving was the most frequent method of 'artistic' display, with color and depth often added utilizing various types of rocks and rock formations. Engravers used tools such as stone picks and sharp flint flakes.

Then painting arrived. What paint materials were used that could stand up so long?

- Red pigment from ochre or iron oxide, the same as that used by the San in southern Africa, was the most durable. Yellow ochre and brown ochre were also utilized.
- Black from charcoal, soot, and minerals like manganese. Red and black were the most frequently used colors.
- White, from kaolin or mica or other materials, was used only rarely.
- Liquids used for mixing: cave water with its natural ingredient of calcium, and vegetable and animal oil as binders.
- Applicators included crayons, fingers, quills, feathers, twigs, leaves, animal hair, thin bones, and pipes made from

bird bones, which the artist used to blow paint on the walls, creating designs utilizing stencils.

Upon seeing the Lascaux cave art in 1941, Picasso said, "We have learnt nothing!" In his book, *How Art Made the World*, Nigel Spivey says, "Uncannily (as it must have seemed to him), the prominent animals at Lascaux were bulls—favoured subjects of Picasso, and indeed, featuring in one of his earliest paintings as a boy. Also, some of the animals depicted at Lascaux have their form emphasized in thick black outline. This is also uncannily similar to a pictorial device favoured at one time by Picasso and his post-Impressionist contemporaries . . ."

This should not be surprising with what we are learning about the San and shaman. Almost everyone on earth traces his/her ancestry back to the San, and artistry as we know it originated with the San-shaman. Therefore, embedded within all artists is a little, or sometimes a lot, of the San-shaman. Picasso was from Spain where Ice Age artists' genes had planted deep roots.

Richard Leakey, renowned archaeologist, believes that the cave painting known as "The Sorcerer" at Trois-Fréres in the French Pyrenees is shamanistic in origin. The human-animal hybrids known as therianthropes displayed on the walls of the cave are common in the shamanistic art of, yes, the San of southern Africa.

Of course outside 'artwork' was also produced, but only a few engravings have survived. Some images inside the caves were engraved into the rock to produce bas-reliefs, a favored technique seen frequently as the major civilizations arose thousands of years later.[4] Caves were used for artwork because of their preservation features and because they offered 'access to the spirit-world'. The paintings were never considered complete but were instead 'works in progress', just as the southern African San believed that their paintings were never complete.

It appears that the cave artists worked in an environment similar to that of a modern art studio. Assistants would mix the paints, keep the lamps or torches burning, install the scaffolding, and prepare the paint 'brushes' or applicators. These assistants would probably become painters and artists also.

It was not unusual for the caves that were used for painting to be extremely large; for example, the cave in Rouffignac, France was over six miles long. The people of the time might occasionally live in cave mouths or natural rock shelters but not in caves. The same basic set of animals was painted in European caves for about 20,000 years with little deviation in the images.[5] Here we have cave paintings in Europe still closely resembling southern African San paintings up until about 12,000 years ago when cave painting stopped.

Why would it stop all at once? Twelve thousand years ago was about the time of the end of the period known as the Younger Dryas, ice age conditions that had seemingly reappeared from 'out of the blue', lasting about 1300 years or so. At its conclusion, the climate improved dramatically and quickly. As mentioned, the vast majority of cave paintings were found in southern France and northern Spain. It was probably an extended community of sorts with several nodes. When the climate changed, the community probably decided it was time to move on—to greener pastures.

Some anthropologists think of these different communities in Europe as disparate groups. But that is a consequence of living in our SN-led world which sees and fosters differences among people rather than the commonalities which are much more prevalent. The SN structure produces an SN mind which also acts in a dispersal-oriented manner. Remember, the hierarchal SN divides everything because it wants everything divided, even the people under its control. Most of the minds within the energy field follow suit. The P-IMPACTS generally resist, many maintaining a circular and anti-dispersal attitude and orientation.

An example of SN divisive behavior is the political system in the U.S. in which two political parties, miles apart philosophically, have rigged the system such that a strong cohesive middle cannot form. With its present setup, other serious attempts to form political parties and join the fray are almost impossible. It is a typical SN approach—keep 'new' blood and ideas out, maintaining the status quo.

The End of the Ice

As the giant ice sheets gradually began to melt about 18,000 years ago in present-day Europe, the rivers became full, some emptying into what is today the Black Sea just north of modern-day Turkey. The area around the lake became a 'Garden of Eden' with extensive animal herds and an abundance of fruits and vegetables. The land to the west of the Black Sea (eastern Europe) was full of thousands of hot springs that provided relaxation and health benefits. The volcanic Carpathian Mountains in eastern Europe were rich in obsidian and flint for hunting tools. Salt and obsidian were also plentiful south of the lake. Why was salt important? Because man would soon learn that salt could preserve meat and thus free time and energy for other pursuits.

You can see where this is very much a replication of early San environments. It would attract San-like descendants like a magnet. The herds of mammoth that humans had depended on for so long were diminishing in number, due partly to the skills of the hunters and to the increasing human population. Modern humans were learning how to survive and multiply, growing from perhaps only 50,000 people on the entire planet 50,000 years ago to over 5,000,000 worldwide 10,000 years ago. Certain areas had probably reached a tipping point where population pressures created a need for a different way of feeding people. The prokaryotes had laid the foundation for the eukaryotic structure that was about to emerge.

The Ascent of Agriculture, the SN, and Modern Human II

We asked the question earlier—how long would the stasis of the San group stay intact as it traveled around the globe? What would be powerful enough to puncture the equilibrium?

We mentioned in the previous chapter that an SN of sorts appears to have arisen in western Europe during the Upper Paleolithic about 40,000 years ago, but it would not have been able to muster a great deal of power because the population was still relatively small. If it asserted too much power, it would destroy the cohesion of the IMPACTS, and then it would have nothing. Today the SN is firmly in control because it can gobble up huge nodes of IMPACTS. At that time, more strength may have rested with the IMPACTS side than with the SN side, or at least it was an even fight.

The onset of agriculture in the Fertile Crescent of the Middle East about 10,000 years ago would be the variable that would truly puncture the equilibrium and introduce the period known to anthropologists as the Neolithic Age. With the cultivation of wheat and other grains, a year's supply of food for a family could be harvested in a few weeks, thus allowing for more-permanent settlements and more of a division of labor. Building houses and other structures would attract some, others could teach the young, rudimentary medicine was an option, and others might be part-time farmers and craftspeople. Roles would be created as needed. Specialization would become the norm. But again, this specialization had its basis among San-shamans, one of whom might be the shaman for rain, another for hunting, and still another for healing.

Agriculture brought about social developments that would lead to a stratification of society and the development of Modern Human II, who was opposite to Modern Human I in many ways. MHII was more male, MHI more female. MHII naturally fit in better with the developing SN. The SN was building up its side of the equation just as black holes had done after stars and galaxies had started forming.

All of this is following the same model we have seen previously. When the eukaryotic cell formed, the nucleus 'brought in', or accreted, prokaryotic bacteria to be its supplier of energy. The nucleus took charge and divided up duties within the cell. The same was occurring in human society, including the specialization of roles. It is the same all over the universe: the creative-formative-productive energy lays the foundation, and then a nucleus forms and takes control.

If the SN were to grow into the force it is today, it would need supporters and lots of them. But innovation generally comes from the anti-status quo, more female side of the equation, so the SN would have to assimilate the IMPACTS somehow. Plus, San-shamans had been natural leaders, and therefore many of the IMPACTS would be the same. Getting these MHI IMPACTS 'tamed' would be a large part of the story. The SN would want to pick the 'good fruit' and throw the 'rotten apples' away.

Understanding the development of this stratification will help us see how IMPACTS energy operates in today's world. In contemporary society, Modern Human I and Modern Human II are very much intertwined, just as the human brain's creative right side is intertwined with the masculine left side, and the atom's nucleus is intertwined with the electrons.

In the first chapter, I told the story of the mealybugs and the herdsman ants in the rainforests of Malaysia. In their symbiotic relationship, the mealybugs provide food for the ants, which in turn protect and care for them. That has also been the story of human beings since the emergence of agriculture and the SN—the IMPACTS have been providing the 'nourishment' for the SN. But our world is not nearly as peaceful as is that of the ants and mealybugs. Agriculture brought a more reliable food supply, but it also brought the male SN and its seemingly insatiable appetite for power and control.

The SN formed because the IMPACTS energy had become more settled, or captured, just as a proton becomes the nucleus when it captures an electron. Following the template of the universe, a nucleated, hierarchal entity emerges to 'manage' the surplus creative-formative-production of the IMPACTS. SN male energy can be aggressive, territorial, and expansionist—just like galaxies and their black holes. Therefore, war became a recurring theme, and remains a popular approach today. Much of it is undoubtedly due to an excess of testosterone. It sounds ridiculous but we can be certain that it is a major contributing factor.

If the SNs around the world truly wanted peace, they could have it. All they would have to do is round up a group of IMPACTS and give them the task of discovering how to achieve it. But real peace is based on justice, and that word is not generally in the vocabulary of the SN, certainly not in the vocabulary of the extreme SNs. Have you heard any SN leaders use the word recently? I haven't. Therefore, unfortunately, wars of resistance and liberation are often necessary to overcome the oppression of a power-hungry SN. The world is continually held hostage by a few extreme SN people, almost always males.

Just as rape is not about sex, wars of aggression are usually not about grievances between (among) people. They are about power and control, the same things that drive a man to rape. That is how unmitigated male energy usually resolves disputes—through force.

IMPACTS and their energy are generally the opposite—peaceful conflict-resolution is usually their preferred approach. But many SNs do not subscribe to it because it 'feels like' weakness and not power. Extreme SNs have to be 'over' others, not beside them.

IMPACTS Cluster

To develop an accurate conception of the development of human civilization, we must understand as much as is possible about all of the different variables at work. Up to now, human history has been mostly a linear look at events over time with little attention paid to dynamics other than the quest for power and control. That would be similar to describing a marriage as a series of arguments without attempting to understand the dynamics of the relationship.

What accounted for the significant increase in population and for the exponential increases that were to follow? The human race now had embedded problem-solvers, and it had reliable people to deliver the solutions. The pre-agricultural San-like people—the bacteria—were morphing into the mitochondrial IMPACTS. The IMPACTS would supply the innovative energy needed for the eukaryotic 'human civilization cell' and its structural-nucleus, the SN. During the long run up until agricultural development, freedom was the word for the San tribe and their descendants who traveled the globe. But the eukaryotic society that was emerging with its structural-nucleus was going to severely limit that long-established independence.

Energy flow as the SN solidified its development about 6,000 years ago. It is the same basic model in use today except that the spirit world is now mostly organized religion.

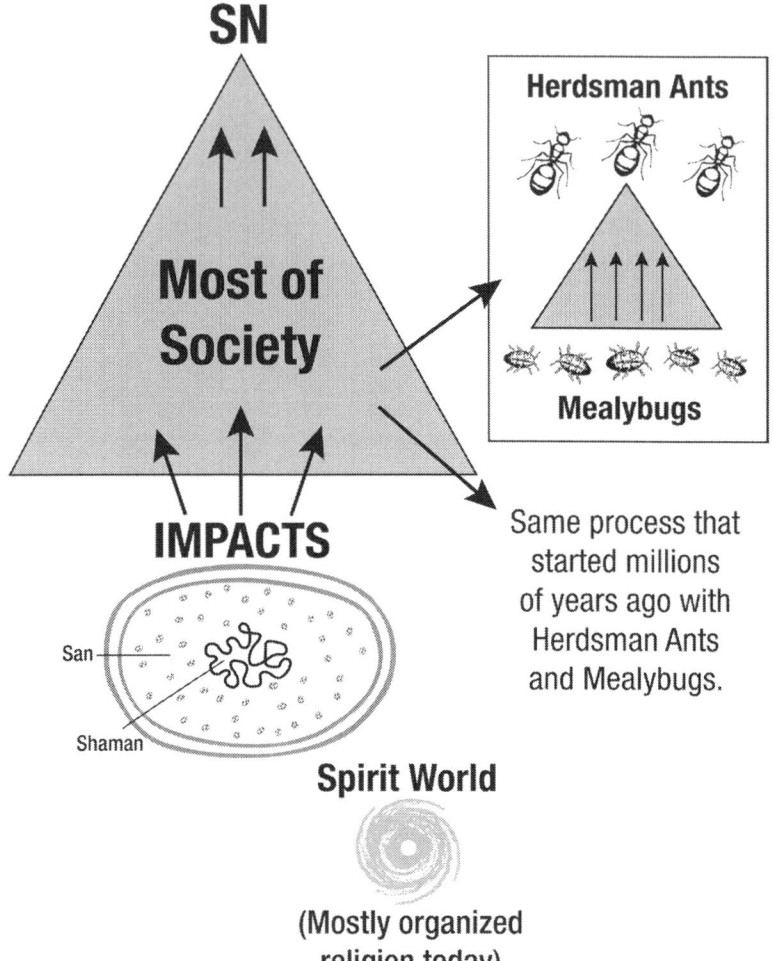

The SN is accreting from the IMPACTS and all other CFPE such as nature, women, and the spiritual realm. The SN intent is to capture all CFPE (IMPACTS energy) in order to strengthen itself and exert power and control. IMPACTS have the same philosophy and attitudes of the San and San-shaman. They provide innovative energy, foundational support, and primary bonding for human society.

As IMPACTS genes spread, so too did the population. They went hand in hand. You could not have a burgeoning population without a burgeoning number of IMPACTS. It was not possible. Today, it is less critical because of modern technology and transportation systems. The IMPACTS don't have to be in the neighborhood any more; with internet and phone connections, 10,000 miles away is the same as being in the next room. And with the machinery and technology that the IMPACTS continually invent and develop, they are in effect duplicating IMPACTS energy exponentially. This allows the SN to continually expand even as the numbers of IMPACTS may actually shrink as a percentage of the population. You can see why it becomes critically important for SNs to foster good relations with IMPACTS-laden countries.

As I said earlier, I believe that the profile of the IMPACTS became established in the human genome long ago when the problem-solving San-shaman became part of the genome along with his support staff, the remainder of the San tribe. If there are significant pressures for the inclusion of certain traits in the genome over an extended period of time, nature seems to find a way. I presume it is further manifestation of what might be the purpose of life itself—to find the best way in the current environment to stop dispersal, and to make it as permanent as possible. That is what it appears the anti-dispersal forces want to do—make the bonding permanent. In our present world, this is illustrated by the IMPACTS' efforts to develop strong institutions that protect humanity and the earth and her creatures.

Inevitably, IMPACTS would cluster. That is what creative-formative-productive energy, IMPACTS energy, appears to do if it becomes stationary—it becomes concentrated. There again, that is what we would expect of an energy that attempts to defy dispersal. Dispersal forces do the opposite—they try to break down concentrated energy.

When IMPACTS cluster in the human world, innovation, complexity, and productivity increase. It is the valence electron phenomenon again. Where you have valence electrons of several different atoms connecting, you can get some very complex molecules of concentrated energy.

Star formation is similar. Stars form in clusters as many will develop near the edge (periphery) of a huge cloud of molecular hydrogen gas when it is disturbed by turbulence, such as from a supernova. Then the stars will create more turbulence which will cause more stars to be formed, and on and on. The creative-formative-productive energy transforms that turbulence and discord into the production of energy, just as the shaman did. From where does the CFPE emanate? I think it comes from the electrons.

There are two major forms of star clusters, globular and open. Globular clusters are huge, sometimes composed of a million stars, and are found in the galactic halo, orbiting the center of the galaxy like a satellite. Open clusters are much smaller, with usually only a few hundred stars, and they are formed in the galactic disk. Open clusters generally disperse after 50 million years or so while globular clusters may contain stars billions of years old. Our sun was probably once part of an open cluster but 'struck out on its own' long ago.

Stars in an open cluster move away from each other according to the 2nd Law of Thermodynamics because there are not strong gravitational forces holding them together as there are with globulars. Even in a globular cluster, peripheral stars will 'escape' from the cluster and either become part of the galaxy or move beyond it. This causes the globular to become even more tightly bound within. It is believed that eventually the globulars, as they lose more and more stars, will collapse and form a black hole.

It is the same in the world of people. If the peripheral-oriented IMPACTS leave the company or organization or country or even family, the remaining entity may become more tightly bound. But that may not be a positive as innovation and diversified viewpoints will be lost. If a significant number of IMPACTS leave with their productive capabilities and energy, collapse may occur. Keep in mind that production prevents collapse. It is true within a star and it is true within a human group.

Clustering of creative-formative-productive IMPACTS energy appears to be an antidote to the 2nd Law. The reaction to the 'threat' of dispersal or to the instigation of dispersal may ultimately be the production of energy, or the preservation of it in cohesive form. IMPACTS people interact with the world in the same way. They try to keep human beings, institutions, and the earth itself 'well', balanced, harmonious, and bonded, which of course were the same goals of the San-shaman.

The First Locations for Agriculture

It was about 11,500 years ago that the oldest known 'temple' in the world, and perhaps the first example of architecture, was built at Gobekli Tepe in today's Urfa Province in southeastern Turkey, an area drained by the Tigris and Euphrates Rivers. It appears that when the temple was built, the workers were still hunter-gatherers. Some believe that this region is the Garden of Eden in the Bible. Nearby communities are mentioned in the Bible, and Abraham was believed to have been born in this vicinity.

The limestone hills and volcanic basalt of Gobekli Tepe also sit near the volcanic Karacadag Mountains, the area that has been identified as the likely origin of einkorn wheat, first domesticated over 10,000 years ago. The hot, dry summers and the wet winters were perfect for farming on the fertile tableland around Gobekli Tepe. As Nigel Spivey mentions in *How Art Made the World*, architecture and agriculture may have developed about the same time as the artisans and others who gathered to work at the site would have required a dependable source of food. Then, upon completion of their intermittent work, they might have returned home with samples of wheat which would have been planted across the area.[6] That would have been the usual IMPACTS way of doing things—aggregating, forming and renewing bonds, producing something of value, and then taking and applying any new ideas back home. That is the way it was done with the San tribes and their shamans, and it has been carried forward with their descendants and the IMPACTS.

You will notice that when the IMPACTS disperse, they are taking practical ideas with them that will help build stronger bonds in their communities or other communities. The SN sees dispersal in a totally different light, one without bonds or with bonds that favor the SN.

When you think about it, that is the way it works with bacteria too. They 'congregate' also, exchange DNA if new adaptations to the environment are needed and available, and then the DNA disperses near and far, changing the community. Creative-formative-productive energy appears to behave in the same manner no matter where it is found.

Just as the San had done tens of thousands of years before, the people at Gobekli Tepe worked extensively with limestone. Their temples were round, a sign of the lingering influence of the San and their circular view of reality, life, and community. Some of the limestone bedrock was carved into, creating a subterranean feel in parts of the structure. Numerous T-shaped limestone pillars, some twenty feet tall and weighing 50 tons or more, were used for support.[7] The same basic model was used in Catal Huyuk, Anatolia (Turkey) 3,000-4,000 years later, and at Stonehenge in England, 7,000 years later.

What does that tell us? It tells us that the same construction methods were spread near and far, probably by emigration. But it also tells us that the genes and the skills locked up within them stayed mostly intact for many thousands of years. The population would have remained very San-like until the SN intruded. As the SN developed, the IMPACTS would try in many instances to maintain their mostly San way of life by escaping to peripheral areas. But they would inevitably be caught and brought under the control of the SN. Sounds like taking a prisoner captive, doesn't it? In a very real sense it was, just as the mealybugs were captured by the herdsman ants, and just as stars are captured by the black hole center of a galaxy.

Animal bas-reliefs, also seen in European caves, adorned the pillars at Gobekli Tepe: foxes, lions, cattle, wild boars, herons, ducks, scorpions, ants, snakes, and others. Again, this may have been a belief that the pillars contained the potency of the depicted animals and therefore were aiding in the protection of the site. Protection is again a primary part of the creative-formative-productive element; it is value and bonding that is being protected.

There were free-standing sculptures also. The oldest statue ever found, that of a six-foot man made from limestone and whose eyes were made of obsidian, was excavated in the area along with other statues. Similar structures to those built at Gobekli were also found at other nearby communities, including Nevali Cori, where houses were also constructed, some 11,000-13,000 years ago. Some of these houses were built right on top of previous ones. Why? In this area, loved ones were often buried under floors and inside of walls. Building on top of the old may have been a way to maintain bonds with the deceased.

Why would the temple at Gobekli Tepe have been built? It was most likely an aggregation center where the same functions were performed that had taken place earlier in caves and rock shelters—rituals, initiations, and other activities, including healing and ancestral worship. IMPACTS energy is productive-cooperative-bonding energy; SN energy is accreting-divisive-separating energy. So when you see this kind of effort to bring people together, it is usually a sign of IMPACTS and their energy at work. It is the same in today's world as IMPACTS will generally be the participants in seminars, conventions, and other such events, wanting to connect with others and exchange ideas, energy, and anything under the sun that will help improve the situation.

Interestingly, the temple was covered in dirt as agriculture got under way. Why would the people cover the structure with dirt after they had worked so hard to construct it? The answer to that question lies hidden with the San-like IMPACTS who actually did the construction. But the possible explanation is that again, the spirits of their ancestors and loved ones were part of the structure, and therefore burial of the entire building was necessary if the living were leaving the area. In Catal Huyuk homes, when one room was abandoned, the spirit-filled walls were defaced, and the entry way was bricked shut. Often this entry way was very small and close to the floor, indicating that crawling was the only way to enter the room. Neolithic people utilized some previous cave behavior above ground. Catal Huyuk was also near a limestone cave.

The Neolithic

The pre-Neolithic Natufian people of the Levant, the eastern coastal region of the Mediterranean from Turkey to Egypt, may have been the first group to discover the preservation properties of salt. At the time, the Levant was lush and full of game as opposed to the desert it is today. The first permanent Neolithic settlements of the area, such as Jericho just north of the Dead Sea and Catal Huyuk over a thousand miles to the north in Anatolia (Turkey), would necessarily have been close to salt supplies.

Jericho to Catal Huyuk became a regular route for hunters and traders, from one salt area to another. The general time frame we are discussing is about 7,500 BCE to 5,600 BCE.

Let's not forget the importance of obsidian, the volcanic glass that was sharper and smoother than today's surgical scalpels. Not only was salt in plentiful supply near Catal Huyuk in Anatolia, the Hasan Dag volcano provided all the obsidian needed for domestic use and trade, much of it with Jericho for cedar lumber and bitumen. Cedar is rare in that it does not rot, making it especially valuable. Bitumen, a tar-type product, could be used as an adhesive. In Jericho, it was used to place seashells in the eye sockets of human skulls. Lime-plastered skulls, sometimes painted (again!) with red ochre to provide a silky finish, have been found in Catal Huyuk, Jericho, and 'Ain Ghazal in present-day Jordan.

In the Upper Paleolithic, people would bring fragments of animals that they had killed and eaten into the caves and insert them into the 'membrane' of the rock. This was like a pact—you could kill the animal if you showed proper respect and returned its soul to the spirit world. Around the world in many shamanic societies, the soul, or spirit, is believed to reside in the bones, and that includes fish and animals. If care and respect are given to the bones and therefore to the soul, then it is believed that more fish and animals will be born.[8] You can see this as a model for many religions—if you behave properly in your life's activities, you will be rewarded by the spirit world. Bone, soul, and rebirth are a recurring worldwide theme.

In Neolithic homes, parts of animals were often plastered into and onto walls, and infants were sometimes buried inside the walls as well. By plastering and painting over the walls again and again, it was as if the hands-on work was a way to stay in touch with the spirits that now resided just beyond the new 'membrane'.

Remember the significance attached to the 'laying on of hands' in the paintings of the San and those of the Upper Paleolithic cave artists, and the many instances of paintings of hands? Potency was believed to be transferred from the paintings and from the spirits on the other side of the rock membrane through the hands. We see painted hands in much of the artwork of Catal Huyuk.

Half the buildings in Catal Huyuk appear to have had spiritual, or what we might call religious, significance. Seven to ten thousand people lived in Catal Huyuk at its peak from about 7,500 BCE to 5,600 BCE. Along with cultivation of three types of wheat and one of barley, they also domesticated cattle and goats in addition to hunting wild cattle and deer. The 'living was easy' which allowed them substantial time and energy for artwork and what we call religion. In addition to sculptures and statues, there was an abundance of the 'Goddess' figures that were seen in Europe during the height of the last Ice Age. Paintings that suggested an enjoyment and vitality of life have been found extensively throughout the remains of the city. The society appears to have been very much like the San community of old—egalitarian with very few hints of hierarchy.

Catal Huyuk also attracted people from the north, following the great rivers like the Danube down to the present Black Sea, and then overland to Catal Huyuk. We can see how it would have been a mecca for artists as well as a center for spiritual activities and trade.

There is disturbing evidence of human sacrifice at Catal Huyuk, particularly among some young males whose skulls revealed a high incidence of head wounds. Plus, the bones of their arms showed signs of injury.[9] Were they trying to protect themselves with their arms? That leads us back to our prior discussion of autism and its frequency among males, and its possible association with the IMPACTS profile. We saw that brain surgery was performed by the Nazca Indians of Peru 2,000 years ago and wondered if it was related to autism. Now the same questions arise in relation to the young men of Catal Huyuk 7,500 to 9,500 years ago.

Our society deals with autism and 'psychiatric' problems by mostly trying to keep them out of sight. Did these earlier societies have even worse ways of dealing with such issues? What could possibly have been the reason for these head wounds? If they were administered by someone, who might that have been?

The present-day Black Sea was, at the time of the early Neolithic, a freshwater lake surrounded by plains and game. As the ice sheets melted, the levels of the oceans and seas rose around the world. Then, about 7,600 years ago, it is believed that the Mediterranean broke through the Bosporus Strait, filling the freshwater lake with saltwater and raising its water level dramatically. Thousands of people around the lake were forced to move in different directions—some westward into Europe, others south to the Levant and the Nile River basin, others northward to present-day Russia and beyond, still more southeastward to Mesopotamia and modern-day Iran. But moving was nothing new to these resourceful people whose lineage went all the way back to the San tribe. This sudden 'turbulence' and dispersal of IMPACTS energy in all directions likely sowed the seeds for the coming civilizations: Greece, Rome, Persia, Mesopotamia, Egypt, and others. When turbulence occurs, IMPACTS start producing 'stars'.

In the next chapter, we will take a look at the development of the structural-nucleus (SN) and the hierarchal society as agriculture turned everything upside down. The mostly prokaryotic way of life gave way to the eukaryotic society.

Summary

By all indications, it was a San tribe that left Africa 60,000 to 80,000 years ago on the journey that ultimately colonized the world. The San could travel out of Africa with confidence because they had already survived for 50,000 years or more using the model of a close-knit, cooperative, loving group with the innovative shaman at the center. Fifty percent of a San tribe might be shamans.

Geography and topography, the natural living environments, are extremely important when discussing the development of modern humans. It appears that in their travels, early modern humans tried to replicate as much as possible the 'primal' environment of the Great Rift Valley of Africa and nearby coastlines. They were attracted to volcanic areas with natural springs and obsidian, plains with rich game, lakes and rivers, a temperate environment, valleys and steep, rocky cliffs, coastal waters where they could find food, and caves.

Around the world are signs of San behavior: similar paintings and materials and techniques, figurines that depict steatopygia, the almost ubiquitous use of red ochre, the utilization of similar seashells in personal jewelry, and much more.

I call the San and shaman community and their descendants who still possess the San philosophy and attitude Modern Human I. MHI is more people-, female-, and solutions-oriented. These are the IMPACTS though a high percentage are now 'mitochondrial'

IMPACTS (SN-IMPACTS) working in close association with the SN, though not always enthusiastic about doing so. Modern Human II developed alongside the SN and agriculture. MHII is more male and tends to fit comfortably in the hierarchal structure where the individual is emphasized and where acquisition is encouraged. This acquisition can be of a territorial nature or simply possessions.

A high concentration of IMPACTS in a given area, such as that around the Black Sea 8,000 to 9,000 years ago, will produce extensive art, culture, and innovation. IMPACTS cluster as concentrated energy, thus forming bonds that strengthen society and prevent dispersal.

Agriculture, which enabled the development of the SN, appears to have arisen about 10,000 years ago near a volcanic region of the Fertile Crescent in the Middle East. Agriculture and architecture, as evidenced by the temple at Gobekli Tepe, may have arisen together.

Footnotes

[1] *The Mind in the Cave*, p. 263.
[2] *The Mind in the Cave*, p. 224.
[3] *The Mind in the Cave*, p. 224.
[4] *The Mind in the Cave*, p. 28.
[5] *The Mind in the Cave*, p. 268.
[6] *Inside the Neolithic Mind*, p. 33.
[7] *Inside the Neolithic Mind*, p. 31.
[8] *The Mind in the Cave*, p. 253.
[9] *Inside the Neolithic Mind*, p. 81.

Chapter 7

The Ascent of the SN

We saw social conflict and tension in the Upper Paleolithic era about 40,000 years ago in western Europe during a time of cultural expansion. This appears to be the natural consequence of sedentary, clustering IMPACTS energy, if there is masculine energy nearby. An SN entity emerges to take control of the surplus production created by the IMPACTS, if given the opportunity. Why don't the IMPACTS take control of it? In today's world they sometimes do because of the SN-IMPACTS hybrid energy we mentioned earlier. Bill Gates is a good example. Linus Torvalds, who started Linux, is a typical example of a P-IMPACTS person who creates, takes control of development, and produces, but is not 'territorial' about his creation, as opposed to Gates. But there are really two forms of territorial—an SN-territorial that tries to control and an IMPACTS-territorial that tries to protect against the SN.

Forty thousand years ago, the lines were probably more clearly demarcated between the small SN that might have arisen and the more established IMPACTS. The SN demands a compliant IMPACTS contingent just as herdsman ants demand compliant mealybugs. The IMPACTS at the time may not have been so yielding. Therefore, tension and conflict resulted. Or the tension and conflict could have resulted from what we see in the modern world—SNs vying against one another with 'my IMPACTS versus your IMPACTS'.

At different locations in the heavily-volcanic Aegean Sea area over 3,500 years ago, there was extensive clustering of IMPACTS but very little conflict and tension within the societies, or between the societies. But these societies were not landlocked as were Upper Paleolithic groups in western Europe, especially during periods of ice. The Aegean civilizations were very mobile with instant access to the sea, which gave them an outlet for their energy and production. They were not captured and thus were not stationary. This may have helped maintain the IMPACTS character of most of the region, until SN elements arrived from the 'periphery'. This peripheral element is believed by many anthropologists to have originated from landlocked areas north of the Black and Caspian Seas.

We mentioned earlier that even within the San, people who had always been hunter-gatherers, a group broke away about 2,500 years ago in Africa and became pastoralists. They called themselves the Khoekhoen (also Khoikhoi), which means 'the real people' or 'men of men', and also 'we people with domestic animals'. The Khoekhoen were the first to call their ancestors the San, which basically means 'people with nothing', terminology with hints of derision and class assignment.

Along with the aforementioned condescension, it did not take long for a basic hierarchy to form within the Khoekhoen based on the number of cattle owned, or wealth. The 'rich' man became the leader of the group. So the elements for hierarchy were there in modern humans from the beginning though the nomadic, possession-free lifestyle provided no space for it. Hierarchy in an environment of scarce resources can be a distinct liability rather than an asset for group survival. When people stop moving, however, hierarchy usually sets in, though it can be benign, or it can be something else.

It appears that the hierarchal model, if the Khoekhoen are a reliable guide, is based primarily on accumulation of possessions, and the contemporary world is strong validation. Even the heavens prove the point as bigger galaxies devour smaller ones. The galaxy with the most matter (stars) rules if the galaxies happen to intersect. But as we have repeatedly seen, significant numbers of IMPACTS have little interest in acquiring property and possessions. Extensive acquisition can be thought of as a form of accretion (male energy), and IMPACTS (mostly female energy) are usually more interested in the opposite, which is giving and producing. But again, today's world is a hybrid of both.

Recently, astronomers observed a super-massive black hole of one galaxy blasting another nearby galaxy with a powerful jet of particles. The jet is about 1,000 light-years across and extends outward 1-2 million light-years, though the galaxies are only 20,000 light-years apart. The galaxies are circling one another and appear to be in the process of merging. Images show that the jet from the larger galaxy has dissipated some of the energy of the smaller galaxy.

Neil Tyson, director of the Hayden Planetarium in New York, said, "Black holes are famous for wreaking havoc on their environment. This particular black hole is disrupting its local region by dining on matter that wanders too close—which is the source of the energy for this jet . . . it is like a black hole bully, punching the nose of a passing galaxy."

It is hard to miss the obvious—the cosmos has the same battles throughout, including here on Earth. The black hole at the center of the galaxy accretes matter and uses it to produce 'bad' energy; the stars use matter to produce 'good' energy. Human beings are just a microcosm of cosmic forces. We have our black holes and we have our stars.

Let's not forget that modern humans are primates just like our chimpanzee cousins, and therefore, we most likely have embedded within us a potential for primate social strategies. Chimpanzees have been known to attack neighbors that they feel are a threat to them and their resources, and then devour their young. Of course we like to think we are above that, but the only difference appears to be that we do not eat them. We put a bow on some of our ruthlessness to hide the chimp-like behavior. Some would say we are actually worse since we now have access to such horrific weaponry, which we can use to basically incinerate people. One reason the San exhibited behavior so 'un-primate-like' was because they were mostly a 'female' group.

Normal primate alpha strategies would not have worked in the unforgiving environment that existed when homo sapiens formed. But permanent settlement, with the inevitable accumulation of property and possessions, changes the dynamics of human interaction. The emerging SN would capture the IMPACTS, its suppliers of innovative energy and community foundation, but then it would also need the specific energy of the aforementioned Modern Human II. MHII would be more 'structural' and less 'creative-formative-productive', closer to the SN, more amenable to its way of thinking, and more individualistic and less community-oriented. MHII would of course be more patriotic and warrior-oriented than Modern Human I because MHI, the

IMPACTS profile that developed from the San tribe and its shaman, formed when war was unknown.

Modern Human I, pleasantly assertive, skillful, compassionate, sharing, and bred to live in a circle, would sometimes have a hard time fitting in with an SN-controlled hierarchal structure. But someone was needed to bridge the peaceful, innovative world of the San-like humans with the newly emerging hierarchal world, and the personality profile of the San-shaman, in particular, would be the only choice.

Part of the role of the San-shaman had been connecting the everyday world and the spirit world. Another type of world was emerging but the process would be the same—find solutions to challenges and then deliver the answers along with results, facilitating transformation. In his original role with the San tribe, the San-shaman attempted to institute balance and wellness. Balance was not a concern with the new SN but wellness of the edifice and the SN leader was. Optimization of advantage was the goal, which often meant the elimination of potential among adversaries. That is the 2nd Law of Thermodynamics, and that means dispersal.

Those who had been facilitators in the San world would be facilitators in the new SN world—if they could fit in with the dictates of the new authority structure. Those who were the travelers in the old world would be the travelers in the new one. The healers in the old one would form the foundation for medicine in the new one. Those who searched for answers in the old world would make discoveries in the new one. That is how it worked—the old world merged with the new world. The new one needed all the roles within the San community structure, and in addition it needed warriors, laborers, bureaucrats, scribes, and servants. And it needed a pliable public. A hierarchal structure demands a significant amount of capitulation and ego sacrifice. It reorients the energy flow. The trick for the SN would be to corral the natural service and innovative mentality of the San-like people, the IMPACTS. That would be easier to do in the case of the San personality than it would in the case of the San-shaman profile.

A society was emerging in which one's place in it would be largely determined by whether he or she produced or accreted. It is the same in the atom, isn't it? The San-like genes (IMPACTS) would create, form, produce, and deliver. Those people closest to the SN would generally accrete. This new societal setup would not have been the first choice of the IMPACTS, but they would do what the valence electron does—try to make the best of the unbalanced situation. Civilization, as it is called, is the result of that effort.

Chemical bonds prevent the 'stuff' in the world from dispersing according to the 2nd Law of Thermodynamics, and chemical bonds are formed by valence electrons. The IMPACTS provide the same bonding 'service' in the human world. Without them, humanity itself would collapse into terminal warfare and oblivion—total dispersal.

Activation energy is the energy required to break chemical bonds in the physical world. The bonds of the IMPACTS that hold society together can also be broken if a certain kind and amount of energy is applied. An example is the Vietnam War where the pressures within the U.S. almost destroyed the country. Actually, that is what warring countries are trying to do—they are trying to break the bonds that hold their adversaries together. The North Vietnamese hoped that their efforts would break the bonds in the U.S., while the U.S. attempted to dissolve the bonds within the North, among the so-called Viet Cong (National Liberation Front) in the South, and between the North and the Viet Cong.

SN destructive energy has produced dark ages at various times around the globe; e.g., in the Aegean about 3,000 years ago and in Europe during the Middle Ages. But it is the same at any time in history, including today—the bonds that hold everything together can be torn asunder. The financial crisis of late 2008-2009 is a vivid example of what happens when IMPACTS energy is not sufficiently involved in day-to-day oversight of almost any human endeavor. Extreme male energy is always waiting to pounce and initiate dispersal. And it certainly did so in this instance.

Look at some of the dispersal that occurred recently: thousands of families lost their homes, millions of people saw their retirement accounts decrease significantly in value, millions lost jobs and businesses, and incalculable opportunity—and trust—evaporated. This is an excellent example of what happens when a structure steadily dilutes its foundational IMPACTS energy. If IMPACTS bonding is not given the proper attention, results are unpredictable but problems are a certainty. The IMPACTS try to make sure that things are done correctly and that protective bonds remain in place. You might call them quality-control specialists.

The financial fiasco illustrates what happens continually in the real world. The IMPACTS often get squeezed out of the board rooms and regulatory agencies because they believe in solving problems and preventing problems through a transparent, hands-on approach. This approach threatens power-control types, whether in politics or in business.

The Hierarchy and the IMPACTS

If a society does not have significant amounts of IMPACTS energy, it will struggle. But unfortunately, a society that is more weighted towards the circle and IMPACTS principles can find itself vulnerable to societies more oriented towards the SN hierarchal arrangement. It has happened innumerable times throughout recorded history, and it is a very real and constant danger today.

Recorded history is a brutal game of natural selection with the SN doing the selecting. The countries, civilizations, leaders, and empires that we read about are generally the ones that, unbeknownst to themselves, learned how to utilize the special energy of San-like people—IMPACTS energy—to advance their agendas. The Egyptian civilization is one example with its remarkable pyramids and other great works. Often, SN leaders used the IMPACTS energy within their domain to capture IMPACTS energy from other societies. But that is the basic nature of an SN, continually reaching for more concentrated potential (energy) for itself and dispersal of potential for its adversaries. You see it in business, domestic and international politics, sports teams, and most other human endeavors. The process exists on a continuum—from genial to ruthless.

As the structural-nucleus (SN) would learn who the loyal IMPACTS were, it would assign them to protected, critical positions; e.g., doctor to the ruling family, teacher or special tutor of the court's children, scribe and keeper of the finances, trusted trade representative, architect, construction leader, stoneworker, special events planner, cultural advisor, mechanical engineer for staged productions, decorator of the living quarters, artist and sculptor for the throne, personal jeweler, weapons designer and metallurgist, military leader, logistician, hunting guide, caretaker of prized animals, craftsperson, and others. The personal needs of SN leaders could get pretty extravagant—pyramids, gold ornaments, palaces, tombs, and sculptures as tall as buildings, just to mention a few examples.

Because these people would be invaluable to the SN structural leaders, they would generally be kept out of harm's way. In that way, their IMPACTS genes would survive and become more entrenched in the gene pool, continually laying the foundation for advancements in every category of emerging civilizations.

If you look closely, you can see what is happening here. A small group at the top is determining how civilization 'advances' and who advances with it. The ones who will move forward are the ones who help the SN move forward. The SN leaders can 'afford' to lose 40,000 soldiers in a day, but they can't get along without their tomb builders or personal doctors. The leaders and their egos were firmly in control. The eukaryotic cell had replaced the prokaryotic one, and human interaction had changed irreversibly.

So now if you had one egomaniacal leader, the society nearby would need one of its own to defend against the first one. Thus, the cycle was begun, and it is still going strong today. The military element was the number one concern for early SNs as 'extreme male energy' was unleashed upon what had been mostly a creative-formative-productive 'female' world. Humankind became threatened, and continues to be threatened, because of the whims of SN leaders.

Often, a hierarchal leader would exhibit the SN element and the IMPACTS element, which is perfectly natural because that was what the world was becoming, a mixture of both forces. These 'hybrid-types', such as Alexander the Great, could be especially successful in attaining their goals, though their goals might appear ruthless to others, and actually be ruthless. But the SN was changing the standards. Wholesale murder could be cast as 'the defense of the throne', and the murderers-leaders could usually escape justice. It is not much different from today, is it?

Humanity itself was becoming more like a theatrical production that was only loosely based on the real story. The SN could script the story any way it wanted. The structure was developing away from the IMPACTS foundation of the original

IMPACTS, the San and shaman. The original structure was pure IMPACTS energy—innovative, cooperative, harmonious, and productive. It was real. The new one, the eukaryotic version, would be quite different—hierarchal, divisive, and competitive with an energy flow to the top. It was scripted and coercive.

The IMPACTS part of the new hybrid leader attracted the 'people'—the SN part appealed to those looking for 'strong leadership'. Modern politicians have learned how to play this game though their disingenuousness is usually plain to see.

Clash of Cultures

As we know, recorded history is a clash of cultures, but these cultures are developed internally through the tension between the creative-formative-productive IMPACTS and the power-control-oriented SN. The same occurs with the atom—it is a structure created by the tension between the nucleus and the electron(s), as is a molecule. Almost everything—or maybe it is everything—in the universe seems to be constructed as a result of the conflicting relationship between these two basic forces, dispersal and anti-dispersal.

IMPACTS are a bottom-up force. That is one reason why they identify with the underdog. This can be a particularly touchy situation for the hierarchal structure, which generally cares little for those who are not contributing their 'fair share'.

The two forces, the IMPACTS and the SN, would have to learn to live with each other—not an easy task. One force generally wanted to expand and assume power and control over people and resources. The other force was more concerned with community and people issues, including the creation and delivery of needed goods and services, what we would refer to as an economy in today's world. But that was just what the SN doctor ordered. Now these 'careful deliverers of quality goods, services, and information' could truly be utilized to strengthen the power and control of the SN by becoming its benefactors through trade and ongoing innovation. It was a 'great' symbiotic marriage. The IMPACTS loved to deliver solutions and cared little about personal reward and recognition, and the SN loved the power that came with the solutions and heartily accepted the rewards and recognition. It is pretty much the same today. The IMPACTS create, form, and produce, and the SN politicians take the credit.

Think about recorded history. Many very unimpressive 'leaders' have managed to be at the helm when great things were accomplished. How do you think that happened? It happened because these people of little talent were the beneficiaries of IMPACTS innovation and productivity. Of course, many leaders did inspire and helped create the conditions required for excellence, but the role of the IMPACTS has not been acknowledged because, quite frankly, nobody knew until now who the IMPACTS were.

The evolving hierarchal structure needed the IMPACTS, but the IMPACTS did not need the SN. It was like the mealybugs and herdsman ants. The bugs did not need the ants but the ants were dead without the bugs. The IMPACTS' solid, self-sufficient personality profile was constructed over tens of thousands of years by the San and shaman group. It could obviously survive on its own. The hierarchal structure developed after the IMPACTS had laid the foundation, just as more complex life forms emerged after bacteria had laid the foundation. Bacteria could survive quite well without the new organisms; new organisms could not survive without bacteria.

The late biologist Stephen Gould, in *Full House: The Spread of Excellence from Plato to Darwin*, makes the same point: "The ingesting animals are just a little blip upon this basic cycle; the biosphere could do very well without them." It was the same with the accretive SN—the IMPACTS didn't need them. But the SN became much more than a little blip; they became a constant source of endangerment—first to human beings and human societies, and then to the earth and all of its life.

No matter where we look after agriculture has taken hold, we will see these two major forces at work, with the SN always assuming the upper hand, sometimes with carrots and sometimes with sticks. But we will also see the creative-formative-production of the IMPACTS and how they have pushed and pulled civilization forward, generally from the periphery, just as change appears to happen throughout the universe.

The IMPACTS will also be going to the periphery to look for 'better opportunities'. The SN of the hierarchal structure will be doing the same, but the motivations will be totally different. The

IMPACTS aspire to innovate, produce, improve—they want to build bonds among people, and thus, they attempt to connect with others on the periphery. They want to leave no one behind as they are borderless and inclusive. The goals of the SN include exerting power and control beyond its current sphere of influence—that is why it will go to the periphery. To acquire the desired resources, it will often have to apply the activation energy required to destroy the IMPACTS bonds that hold another society together.

But here is a tragic irony in the conquest that we alluded to earlier. The SN will use its IMPACTS to supply much of the activation energy needed to break the IMPACTS bonds in another land. So the SN pits the IMPACTS against the IMPACTS, facilitating the destruction of IMPACTS by IMPACTS. That would be analogous to a colony of herdsman ants using the mealybugs under its care to destroy the mealybugs of a neighboring colony. The mealybugs and the IMPACTS are very much alike—they just want to create and produce 'good stuff'. Their business is not destruction. It is creative-formative-production, which is precisely why the SN wants IMPACTS within its own domain and why it wants them destroyed within an adversarial domain. The IMPACTS mean power to whoever possesses them. The more IMPACTS, the more power.

In our entire discussion, the IMPACTS have been the facilitators for good. Of course, that is not always the case, but generally it is true. Now you can see that the SN will not hesitate to use them for nefarious purposes—to enable its exploits. IMPACTS do not like to think of themselves as enablers of destruction, but unfortunately that is often the case as they go about doing what they love to do, which is innovating. The SN goal is to convince everyone that it is us as ONE against them as ONE. In that way, the role of the IMPACTS as enablers can be hidden, even from the IMPACTS themselves.

Underneath all of the obfuscation and theatrics, this is how the modern world operates, especially in warfare, business, and diplomatic circles. "It is my IMPACTS and supporting cast versus your IMPACTS and supporting cast."

Managing Energy

The onset of agriculture occurred as the climate changed dramatically for the better, which created a tremendous increase in available energy on the planet. Where we see that kind of increase, we see dramatic changes occurring. The species homo sapiens sapiens, or anatomically modern humans, is believed to have emerged as the climate improved about 130,000 years ago. The Cambrian Explosion, the incredible burst of life forms about 550 million years ago, also occurred after a long ice age, perhaps even a snowball earth scenario.

From the SN to the IMPACTS: "We want you to rearrange some of your priorities. Yes, we want you to care deeply for your family and community and to continue to innovate and produce, but we want you to do it all within the greater service to the throne (state). We will protect you in exchange for your innovative energy." It was the mealybug-herdsman ant contract, and it was administered the same way—through natural selection.

If you are an SN and you want the benefits of production moving to the top where you have positioned yourself, then you want the most efficient form of organization in place to aid that flow. If you had a big 'commune-type' arrangement as you had with the nomadic San, then too much energy would be spent within each particular group. The concerns of the group would become the concerns of the individual. Therefore, you need something smaller, a nuclear family, with the male force acting as the agent for the SN and the female force supporting the male force, and duplicating. And that is what we have had for the past few thousand years—sublimation of the female force, including IMPACTS males, to the male force, with everyone essentially working for the SN. It is only recently that cracks have appeared in the armor with females breaking free somewhat—in some parts of the world. In other parts, they remain very much imprisoned.

Recorded history, like everything else, is the story of energy. There are those who create and produce, and there are those who accrete and utilize. The 'atoms' and 'molecules' were forming. The productive energy of the valence electron IMPACTS and the molecules formed by them were being captured by the structural-nucleus, the SN. But this was not a balanced situation for the IMPACTS. Their cohesive bonds had been broken, and they would have to keep connecting and connecting ad infinitum in order to try to get a workable balance between themselves and the SN. The effort continues today. The results are what we see all around us as the IMPACTS continually attempt to 'save' the situation.

Here is the conundrum for the IMPACTS. The more they connect with other IMPACTS and the more creative-formative-production ensues from those collaborations, the stronger the SN becomes. It is like the Milky Way—the more stars that are created, the stronger the galaxy.

I recall in the late 1970s and early 1980s that there were serious discussions about instituting a national industrial policy because the country appeared to be falling behind the Japanese. But soon after the brutal recession of the early 1980s, Steven Jobs, Bill Gates, and other innovative IMPACTS initiated a new phase of the computer era, and the U.S. economic situation began to ascend. When the economy is healthy, then the SN and its military are healthy because they are accreting their power from economic production. An unhealthy economy means fewer taxes and less borrowing power, and potentially a loss of standing in the world.

The end of the Cold War and the fall of the Berlin Wall appeared to brighten the prospects for a healthier world, with many predicting a significant peace dividend monetarily. The Mac-PC revolution and the dawn of the Internet age seemed to solidify the economic situation for the U.S., thanks to the innovation of the IMPACTS and the hard work of their supporting cast. This is how it works in the real world; the IMPACTS get things rolling and the SN takes the surplus value, tangible and intangible, from the production. IMPACTS also seem to step forward when their contributions are needed most.

But everything is temporary in today's world. 'Success' is fleeting. It was less than ten years ago that some economists were predicting the end of the business cycle—no more recessions and budget surpluses as far as the eye could see. There was even talk of paying off the national debt. The bombings of the World Trade Center on September 11, 2001 changed the equation immediately, and reminded the world how fast things can unravel. Quickly the U.S. was involved in two wars and red ink flowed once more. SN leaders have a tendency to see very few variables operating in the world, and hence they generally ignore the periphery. But, as we have noted time and again, that is where change lurks.

Another brutal recession is currently underway, and this time President Obama has instituted an industrial policy for automakers. Left to its own devices, most of the U.S. auto industry would have folded. Why couldn't the IMPACTS resurrect the car companies without government help? IMPACTS and IMPACTS energy cluster around the latest cutting-edge industry or industries, and today information technology and electronics are the leaders. IMPACTS energy is not limitless. Quite the contrary—it is a small percentage of the total energy just as the valence electron is a small part of the total energy of the atom. IMPACTS too are a small percentage of the total population. Yes, there are IMPACTS in the auto industry, and when it was in its infancy, it attracted them in clusters. The reinvention of the industry with emphasis on environmental issues will attract more IMPACTS, but will it be enough to revitalize the industry in the U.S. and make it competitive internationally?

The usual sequence of events is this. IMPACTS start the industry and develop it to its maturity. During the process, IMPACTS will start developing natural outgrowths from the nascent industry, creating nodes which will attract IMPACTS in clusters. As maturity is reached with basic know-how and technology in place, IMPACTS will be spreading out to other areas. But they will continue to staff the key positions of the recently matured industry to ensure high-level functioning. The railroad industry is a good example. We can be certain that it attracted IMPACTS in droves in its infancy. But trains opened up all kinds of new industries (nodes), which acted as magnets for IMPACTS. Remember, IMPACTS are 'new-beginnings' people. Railroads are still going strong, helped in a big way by IMPACTS who have a fascination with the industry.

The process always works the same: the IMPACTS start the 'new', and IMPACTS cluster around it. This new industry opens up other new industries or revitalizes existing ones, creating IMPACTS nodes. The IMPACTS are doing what they like to do—creating-forming-producing, and in the process they are building bonds. And the SN is doing what it likes to do—taking the power and control that comes with increased production and bonding.

It will be interesting to see what happens in this particular situation where the usual sequence of events has run its course and the government is attempting to jumpstart the resurrection of an industry that had matured but had not responded to market realities. This reveals that IMPACTS were not involved in decision-making to a significant degree or else there would have been more movement toward an improvement-oriented approach. But one problem for Detroit has always been that the car companies also have to be in the healthcare business. Business should be about the delivery of valuable goods, services, and information—not healthcare. That is a governmental function. Making a profit is difficult enough; businesses do not need additional burdens.

Again, we have seen with the recent financial meltdown what can happen when elements of the SN continually ignore IMPACTS foundational principles, and overreach in their quest for more and more power and control, of which money is a primary ingredient. No amount of IMPACTS energy in the world can save us from the destruction that follows such egregious carelessness and missteps. A significant portion of the SN and its supporters do not like strict regulatory practices, which can be thought of as 'chemical bonds' holding the energy in place. Many want few controls on their actions, and therefore will lobby hard for the dissolution of the protective bonds that society has constructed, or will find ways to circumvent them.

One thing is certain—the IMPACTS will have to put Humpty-Dumpty back together again. As the economy regains some vigor, the SN will resume its normal accretion along with its suspect practices—unless it is 'hindered from doing so'. The meltdown also reveals a recurring theme since the advent of the SN—it takes only a very few to bring the whole structure down. But it takes millions upon millions to put it back together.

A Review of IMPACTS Characteristics (Based on the San and Shaman) and SN Characteristics

So now we have a pretty good template for an understanding of recorded history and present-day society. The foundation has been laid. It is a battle between two major forces, the IMPACTS and the SN. I will be publishing a book soon that will look at selected periods in history. The history book will illustrate the SN versus IMPACTS battle in much more detail, and will show how the dynamics of that battle shaped events and outcomes. Plus, it will reveal how the IMPACTS themselves were and are the manifestation (force-carriers) of IMPACTS energy, a counterforce to dispersal forces. The same forces are at play in today's world.

Some of the questions addressed in the history book will be:

- Why were the Aegean civilizations so advanced 3,500-4,000 years ago? What caused their downfall?
- What really happened in Greece? Was the experiment with democracy a compromise between the SN and the IMPACTS? Why was the Golden Age so brief, and was it really that golden? Why did the rowers of the trireme ships have so much power, though little or no property? Why were the Ionians the real founders of the Greek Civilization as we know it?
- Why were the Etruscans the real, early power behind Rome? Did the Etruscans start the concept of 'hell' as they were about to be *captured* and incorporated by the Romans?
- What caused the Dark Ages in Europe? What did the Catholic Church have to do with it?
- How was Christianity similar to the San philosophy? How was it different? What did Jesus have in common with the San-shaman?
- Was one IMPACTS family the driving force behind the Renaissance that started in Italy? Why did these periods of 'enlightenment' pop up occasionally? Was it because of IMPACTS clustering and the pressure to break free from SN constriction?

- Johannes Kepler, the great scientist who correctly explained, among other things, planetary motion, also had to defend his mother against a charge of witchcraft. Why was that not as strange as it might appear?
- What part did the Protestant-Catholic divide play in the Industrial Revolution? Why did it start so far north, on the periphery, and in Britain? How did so many innovative people end up in one area, and why were so many of them called Dissenters?
- What role did the IMPACTS play in the French Revolution?

Let's review the IMPACTS foundation which is based on the San and shaman. That will help us understand that side of the 'battlefield'.

- Strong people concerns—healthcare, poverty issues, opportunity for all, representation for all, equality, fairness, justice, pulling for the underdog, leaving no one behind.
- Creative-formative-productive energy. This is a big category. It can be childbirth, invention, entrepreneurialism, artistic and cultural pursuits, cutting-edge technological innovation, medical and scientific research, and more. It is all about 'new beginnings', adding something that will improve the existing reality.
- Emphasis on youth and education, and ancestors. This illustrates the importance attached to the connectedness between the old and the new—to the cycle of life.
- Harmony with and a love of nature.
- Delivery of valuable goods, services, and information, and being entrusted with the safekeeping of value. Trade is extremely important, as much for the exchange of ideas and culture as for economic reasons.
- The inter-mixing of cultures. IMPACTS mostly see borders as artificial divides.
- Always looking for opportunities that will enhance the prospects to develop and realize potential, for all concerned. This often means emigration, travel, trade, mining, and exploration—going to the periphery, reaching

beyond the existing boundaries—just like the valence electron.

- The importance of water. As San descendants spread out around the globe, water was their highway, escape route, buffer, and a food supply. With more permanent settlement, it took on added importance: canals and aqueducts were needed, flooding was used as an irrigation tool, and water-powered machines were invented and utilized. IMPACTS have always been very proficient in using water to aid the human community.

- Attempts to duplicate the early San environment. After leaving Africa, the descendants of the San were attracted to topography and geography similar to San environs— rocky terrain, volcanoes, savannahs-steppes, rivers, lakes, coastlines, mountains and deep valleys, caves and rock shelters—with an eye toward independence and privacy.

- Early, consistent reliance on volcanoes. Volcanic areas not only provided obsidian, fertile soil, and natural springs, but they also enabled people to create underground cities out of the surrounding soft tufa rock. These underground habitats would be used extensively as refuges against aggressive SN forces.

- So-called rebellious elements responding to 'unfair' SN demands and practices. Where you find innovation, you find dissatisfaction with the status quo, and you find concern for the less fortunate. It all emanates from the desire to help people get well and stay healthy. That is the base of all of it.

- The development of art. Art and its practical applications originated with the San-shaman and then developed around the world. This eventually took many forms: sculpting, painting, performing (theatre), dancing, singing, jewelry-making, craftsmanship, ceramics and pottery, metalworking, and more. Metalworkers were invaluable to the SN because of their weaponry design and construction skills.

- Ongoing efforts to restore balance which often means taking the situation from unwell to well or resurrecting its potential.

- A circular attitude toward life. Circular architecture and construction were signs of IMPACTS influence as was an emphasis on symmetry in community designs.
- Stone-cutting and construction skills, including the transportation of incredible tonnage over significant distances. The SN would take full advantage of these skills as it began to dominate.

Some SN elements will include the following.

- 'Strong' leader at the top receiving special treatment though talent may be lacking.
- A leader who often believes he is part-god or sent by god or chosen by god, or one who tries to convince the people that he is.
- Heredity is often the template for the transition of power.
- Accretive behavior with attempts to conquer lands, resources, and other people, necessitating a strong military.
- Attempts to 'keep the people in line'. Usually limited representation in the halls of power of the people who actually provide the energy for the functioning of society.
- A clear division of society into classes, principally the producers and the accretors.
- IMPACTS energy and innovation directed toward 'emperor' goals, often of a very personal nature, such as for the pyramids or extravagant tombs or palaces. Later, IMPACTS energy and innovation would be utilized to keep the SN edifice strong and powerful.
- Control of productive energy. IMPACTS energy was originally directed toward the people of the community. The SN redirected it—upward instead of outward. Those who continued to direct it outward could be in danger.

Can the development of human civilization really be reduced to these two major factors? Am I simplifying a very complex problem? No more so than biochemists do as they try to sort out the foundations of life's chemistry, or Einstein did when he reduced matter and energy to a formula. The atom has only two major forces of which we are aware, the proton and the electron, and look at the complex world that has emerged from it. And that world has come about mainly through the connecting actions of a tiny part of the energy contained within, the valence electrons. Nature is efficient. It can do a lot with a little, especially since it appears to have an extremely lofty goal in mind—hindering dispersal. It appears to all come down to two major forces, male and female. The proton is male; the electron, female. Dispersal is male; dispersal-hindering is female.

- Male
- Pulling inward   • High mass
- Power-control oriented
- Energy-dependent   • Status Quo
- Black hole aspects
- Modern Human II characteristics
- Capture-Accrete
- No creative-formative production
- 2nd law of
  Thermodynamics adherent -
  dispersal oriented

SN

IMPACTS

- Female
- Mobile   • Mining
- Forming bonds
- Energy-independent   • Anti-status quo
- Connecting-cooperating-sharing
- Freedom-oriented   • Reaching outward
- Duplicating-replicating   • Stars
- Low mass   • Creative-formative-
- Captured-Accreted   productive energy
- Modern Human I characteristics
- 2nd Law of Thermodynamics
  violator or resistor —
  dispersal hindering

Life is actually based on four major elements or what we could call atomic configurations: carbon, hydrogen, nitrogen, and oxygen with a little phosphorus and sulfur added in. The acronym is CHNOPS. What should we make of the fact that energy is believed to be produced in stars in two primary ways utilizing these four elements? The first is through hydrogen fusion (proton-proton chain reaction), and the second is through the CNO cycle, or carbon-nitrogen-oxygen cycle. It is as if these four elements, or atomic configurations, have their own anti-dispersal, recycling machine in operation. They produce energy in stars and then they utilize that energy to develop and sustain life. All of that could be a happy coincidence, but I doubt it. It might be useful to think of these four elements as the IMPACTS elements. They make things happen in the category of life.

But again, words and names can really confuse the issues. What we are talking about are different concentrations of energy. The same concentrations appear to work together to produce energy and to utilize that energy for life—to hinder dispersal.

We don't look at things in a connected way as the San did because we live in an SN-controlled world that wants to disperse cohesion, including the view of interconnectedness. An emphasis on cause and effect, which the San-shaman had, gets tossed aside. It has become almost archaic as human consciousness continues to take on the prescribed characteristics of the dominant SN, which discounts reasoning and the search for truth in favor of pat answers, manipulation, deception, and opportunism.

Our understanding of the universe is severely limited by the quality of our assessment tools. Does that mean our brains? Our brains are magnificent instruments, but they are undoubtedly limited, and their energy is being directed away from open inquiry toward service to the SN. That redirection is a significant distortion in any quest for 'truth'. That is why most major discoveries about the 'nature of things' are made by people loosely attached to the predominant structure, either physically or psychically, or both.

Humanity has a basic simplicity just as life appears to have. We cannot see it because the SN does not want us to see it. It is not a conscious decision on the part of the SN; that is just the way a nucleated structure works. The nucleus draws in the surrounding energy, reorienting it to fit the needs and desires of the nucleus.

Even with the inhospitable environment, massive stars have been known to form in proximity to a black hole, revealing a tenacity that has surprised astrophysicists. This is what is needed in the people world as well—massive IMPACTS structures to counteract the SN dispersal tendencies.

As long as the world remains nucleated and hierarchal, there will always be an SN that will try to control the creative-formative-productive energy, just as the Church attempted to do during the Middle Ages. The SN reach will continue to extend far and wide just as it has for the past 6,000 years or so. Just the fact that we are having this discussion is a sign that the SN has been truly masterful at not only creating a reality, but also in preventing those captured from seeing the dynamics of how it has been achieved. Einstein's general theory of relativity applies—the vast majority of humanity is firmly in the SN grasp. The SN is the conductor of the symphony, and humanity is dutifully playing along. The periphery, to a very large degree, appears to have been marginalized. But is that really the case?

Appearances can be deceiving, and answers can still undoubtedly be found on the periphery, leading to a more balanced situation. After all, that is what the valence electron is doing—it is resurrecting balance by reaching out to the periphery, and in the process forming a strong molecule. The San-shaman did the same—he reached out, discovered solutions, and instituted balance. It can potentially happen more often within human society and on a much broader scale, but the process must first begin with an awareness of the dynamics involved.

Many will argue with my thinking, saying it is too simplistic. Actually, the current view of how the world and universe work is the simplistic one. It is not that different from the 'earth is the center of the universe' model predominant in pre-Copernican Europe. That model served the Church and its bureaucratic power structure well. The current model does the same—it serves those at the top of the hierarchal structure. The hierarchal structure dictates what it wants and the kind of people that it wants to aid it in its quest for survival and prosperity. It does not want runaway, inquisitive, creative-formative-productive energy. It wants controlled CFPE that will fuel the edifice—just as gasoline fuels a car. Controlled energy is useful; out-of-control energy is not. The

SN usually gets what it wants, through subtle and not so subtle coercion.

"Boost us and you will get some of the goodies" is part of the SN philosophy, and patriotism is utilized as a mechanism to draw energy to its side. The object is to continually manipulate the situation such that the people who are contributing the most to the well-being of the SN are receiving enough benefits so as to be reasonably content. The SN cares little for those at the bottom of the hierarchal ladder. Their opinions do not matter because they have no real political power, though their contributions are essential to the success of the SN. The language of the SN is power and control. To deal with their ilk, you have to speak their language.

Of course there are IMPACTS elements within the SN, and sometimes the IMPACTS actually gain control of the SN, but usually with strong power-control elements of their own. Keep in mind that this is an SN-IMPACTS hybrid world. There is little 'pure' IMPACTS energy; most of it has been captured and embedded.

IMPACTS energy is usually a step behind SN energy, reacting to it rather than taking charge. The SN usually just takes charge no matter what the IMPACTS do, or certainly attempts to do so. But the story appears to be the same in the atom—the nucleus takes and the electron reacts to being taken.

Deir el-Medina

An example of the struggle at work between the SN and the IMPACTS in early civilization is the ancient community of Deir el-Medina, a village in Egypt founded around 1500 BCE. At its peak, it included about 70 houses and a few hundred people. Deir el-Medina was home to workers, and their families, who built and decorated the tombs of the Pharaohs in the Valley of the Kings. These positions were hereditary as jobs passed from father to son. Women, accorded mostly equal status in Egypt, were provided with servants by the government to help with domestic functions.

In the following passage from Nigel Spivey's *How Art Made the World*, hierarchy means the structure of society: ". . . everyone within the Egyptian hierarchy belonged, as it were, to the pharaoh. Somewhere in the lower middle section of the hierarchy were the artisans whose handiwork remains such a conspicuous presence in Egypt to this day. The artisans were not creative individuals at liberty to undertake commissions as they pleased. They lived and worked in groups, and they worked to a preordained set of rules and specifications. There was no value of originality. Art existed to illustrate the cosmic order of things. Conformity was paramount."

In fact, Egyptian art remained basically unchanged for 3,000 years. That shows the clear ability and power of an SN to first, capture creative-formative-productive energy, and then to orient it to areas that will foster SN well-being and supremacy. The Egyptian SN basically froze IMPACTS energy in its tracks because it could control it and make it do what it wanted it to do.

You can see that the template for the modern world was in place. A very small core was in charge, and as Spivey said, everyone belonged to that core. It is the same model we have today, with only slight variations. The energy and production of today's 'artisans', the IMPACTS, also primarily 'belong' to the SN and the edifice that has developed around it. If you are visualizing a galaxy with a black hole at the center, you are getting the picture.

The workers at Deir el-Medina were part of that lower middle class, better educated and better paid than the vast majority of their contemporaries. But here again, they were contributing immense value to the throne; they were therefore getting the goodies. At times, they were joined by foreign workers, just as occurs in today's world as demand for certain skills reaches across borders. Remember, the San knew no borders, so expect to see the same attitude in IMPACTS throughout history. IMPACTS are the ones who have always treated the world as birds do. They go wherever they want without regard to boundaries—unless hindered from doing so.

Contrary to popular belief, there is no evidence that the tombs and pyramids were built with slave labor, but rather by skilled Egyptians and other craftsmen, and perhaps some craftswomen. If you have excellent engineering skills, combined with innovative, tireless workers who are treated well, production can be surprisingly extensive. Think back to the areas we mentioned previously such as Stonehenge, Tiawanaco, Nan Madol, and Machu Picchu, where stones weighing 50 tons and more were transported huge distances, sometimes up steep mountains. How did they do it? The same way the Egyptians did—with incredible ingenuity, cooperation, and skill.

We see positions at Deir el-Medina that will carry over into civilization-building because the people who filled these positions were vitally important to the rulers and their success. Deir el-Medina was a microcosm of what was to come. Scribes kept records of equipment used, supply inventories, payments to workers, and work accomplished. Such people were invaluable. Today, accountants and bookkeepers fulfill a similar role which explains why those jobs are a magnet for large numbers of IMPACTS. IMPACTS make sure everything is done 'just right', which is perfect for a ruler who just wants to play at being a ruler.

To understand human society, all we have to do is look at a tiny cell once more. Everything works in the eukaryotic cell because of the reliability of mitochondria, and it is the same in trillions upon trillions upon trillions of cells, and has been for the past 1.5 billion years or so. An SN survives because of that same reliability from IMPACTS people.

Stone workers, of course, were the mainstay of the operation, and they were urgently needed across the developing world. Hence, their IMPACTS genes were often disseminated far and wide. Other workers at Deir el-Medina included coppersmiths, potters, carpenters, sculptors, draftsmen, and a part-time physician. The workers also had a labor support group which augmented their efforts. These people carried water, fished for food, gardened, cut wood, and performed other necessary functions. Early on, the foremen were not chosen from the workers, but soon it became apparent that one of the workers should be the foreman, working closely with the scribes. How do we know all of this? Records and personal notes have been discovered that add depth to our understanding of the people and the village.

It appears that Deir el-Medina was eventually abandoned around 1100 BCE due to strife and wars, and the workers may have been sold into slavery. In such a case, they would have been captured in a different way and their skills used by another power or group. We can be certain that it happened frequently. Here, the IMPACTS attitude of 'make the best of a bad situation' would come into play again.

This example shows us how civilization was developing. A small group at the top was diverting tremendous amounts of resources and IMPACTS energy to itself, in this case from the lower part of the middle section of society. This clearly enhanced the power and prestige of the rulers. The members of the society could see the mammoth structures and the power that was accrued from them, and would almost certainly have given the credit to the king himself, not to the small core of workers who were actually designing and building the monuments. When we look at history, we do the same. We do the same in today's world as well. We assign successes to rulers who may have little competence. And when they screw up 'royally', we are sometimes reluctant to put the blame squarely on their shoulders. If you look at these last few sentences, you can see that many of us have been 'naturally selected' to put the rulers in the best light, even when they may deserve quite the opposite.

Egyptian leaders were learning that they would have to do certain things in order to have their best chances of success. Highly competent record-keepers were required. Hands-on people in charge of construction should come from the experienced worker corps. The prized workers should be segregated from the rest of society, treated well, including receiving competent healthcare, and given ample freedom. Some of the workers had second jobs, building furniture and burial materials for people in surrounding towns.

It doesn't look that far removed from today's society, does it? And it is not, except for one glaring difference—today's society does not especially value the hands-on worker, even if that worker is making furniture for the Oval Office in the White House, or is installing a cutting-edge security system in the halls of Congress. The SN values the 'upper echelon' people. The ones putting things together are often assigned scant importance. Therefore, in this country, their access to the 'goodies' of society, such as adequate healthcare, is often limited. The SN, though, is masterful at making these workers feel responsible for their lot in life, and turning any anger they may feel about the situation back upon themselves and others. The SN generally accepts little feedback.

Likewise, the U.S. SN is really not part of a feedback loop. It sets the agenda and enforces it, and solicits few opinions. When was the last time you received a questionnaire from the government seeking your views on the healthcare situation in the U.S. and how to improve it? Any questions from the government regarding your opinions about the U.S. course in the Middle East? Those would be actions of a democracy, and what we have is a republic. There is a clear line of demarcation between those who craft the laws and those who provide the energy that enables the entire structure to exist. Tight filters have been constructed to keep these barriers firmly in place. What does the U.S. SN want from you? It wants productivity, taxes, loyalty, acquiescence, deference, and cheerleading—and few opinions.

Dark Energy

For billions of years, the universe was in a creative-formative-productive stage. But long ago the basic structure took hold, and creative-formative-productive agents cannot now affect the structure the way they could in the early going. Star formation has actually peaked. Something that astrophysicists call dark energy assumed control about 5-7 billion years ago after billions of galaxies had formed. Galaxies are like eukaryotic cells—they are hierarchal, nucleated structures. Mitochondria supply cells with energy—stars do the same for galaxies.

Dark energy, to me, appears to be the intensification of the existing dispersal force, a force that cannot exert itself fully until the CFPE has done its job of laying the foundation for the structure. Dispersal appears to occur on two levels as the nucleated structure forms. First, the nucleus breaks up the cohesion of the creative-formative-productive energy, the CFPE or IMPACTS energy, and reorients the flow to itself. Subsequently, the nucleated structure starts moving away from other similar nucleated structures. It happened after the emergence of agriculture as the San-like remnants gave way to the hierarchal SN; they were captured and their energy reoriented. The same occurred with bacteria as they assumed their role as mitochondria within the nucleated cell. As the SNs developed, they quickly started moving away from one another. But still they sometimes collided (war), pitting captured IMPACTS energy against captured IMPACTS energy—just like dueling galaxies.

The same process of structure formation occurs in human endeavors. There is generally a flurry of creative-formative-productive activity, such as you might see with entrepreneurial activity. As the dust settles, a hierarchal structure, a business in this case, emerges along with a structural-nucleus (SN). The SN of the business takes control of the CFPE, the IMPACTS energy, and attempts to manage it in such a manner as to enhance the SN and its success. It also tries to solidify a spot for itself apart from other similar businesses.

It works the same in daily life as the hierarchal 'real world' breaks apart the cohesion of the creative-formative-productive family unit. Soon after birth, the child starts moving away from the mother, and eventually it will come under the sway of the SN-led structure, and will in most cases serve the SN in one form or another. This leaves the family structure weakened and splintered. But then the children of the family will often start their own families, or creative-formative-productive units. These too generally serve as an incubator for the nucleated and hierarchal structure, which again takes the energy from the family for different roles of service to the SN, again weakening the family. But the process continues on with more children and their families. It is as if the universe is set up for the formation of nuclei of different varieties with creative-formative-productive energy, usually in the form of IMPACTS nodes, 'feeding' the nucleus so it and the nucleated structure can become bigger and stronger. The family is an IMPACTS node; so too a business. All are in service to the SN.

Creative-formative-productive energy (CFPE and IMPACTS energy) leads to a managing nucleus that accretes the CFPE. The nucleated structure appears to be a compromise, or hybrid, of the 2nd Law of Thermodynamics (dispersal) and the dispersal-hindering forces.

CFPE (IMPACTS Energy)	Nucleated structure
Electron	Atom
Prokaryotic cell	Eukaryotic cell
Right-brain development during the San	Today's consciousness with left-brain control?
San and shaman	SN and present model
Women	Family
Stars	Galaxy
IMPACTS	Businesses, organizations, IMPACTS nodes
IMPACTS Nodes	Communities, cities, states, countries, civilization

It looks like an impossible double-bind for the creative-formative-productive IMPACTS energy. It enables the formation of a nucleus, which then takes control of the IMPACTS energy, using some of it to maintain and improve the nucleus and its structure, and discarding other parts of it. What is the IMPACTS energy to do? It is truly captured. It must be the same feeling that a mother often has as she sends a child off to war in a faraway land. The structural-nucleus is taking what it wants of the creative-formative-productive energy in order to strengthen itself—theoretically. Actually, it is just accreting the energy needed in order to fulfill its own wishes. It cares little for the person who possesses the energy that it takes. Regretfully, the soldiers are not serving the country; they are serving the SN edifice.

Most astronomers believe that because the phenomenon called dark energy is in control, the universe is inexorably headed towards a future when black holes will rule. But even this stage will not be the end, according to the proposed scenario. Eventually everything, including all of the black holes and all matter, will be pulled or pushed apart, and the 2nd Law of Thermodynamics will have been realized—everything will have been dispersed equally throughout. But as we are learning, change resides on the periphery, and to my knowledge, we have no understanding or conception of the periphery of our universe. Some astronomers think there may be parallel universes. If so, then we have something on our periphery. So I would hold my bets. You never know what is 'out there'.

The Big Bang could actually have been the culmination of the 2nd Law of Thermodynamics in reverse. Some have speculated that Big Bangs occur repeatedly as the universe expands and then contracts.

Is human civilization headed for a black hole era? Are we in its formative stages, or are we already there? The world certainly shows signs of it as various SNs around the globe attempt to accrete all available IMPACTS energy, stretching and fracturing the bonds that hold humanity together.

Change

You see the struggle between SN-structural energy and creative-formative-productive energy (IMPACTS energy) everywhere you look—in business, in communities, in families, in politics, in your own mind. As I am writing, my CFPE is working with the structure that I am creating, a book in this instance. A paragraph becomes part of the structure, but then my creative IMPACTS energy wants to change it, and so the structure gets altered. The structural energy of the book is very amenable to my creative-formative-productive energy because I am the only one involved; I am the structural energy and the IMPACTS energy. Eventually however the structure is going to prevail—a book is going to be printed. Then it will become far less amenable to change. That is the way the real world works. Structures are not easily changed after the creative-formative-productive energy has done its job.

One reason IMPACTS generally love the worldwide web is because it is more like a colony of prokaryotic bacteria than it is a nucleated structure. Information travels quickly and so too can 'DNA'; e.g., alterations to the code for open-source Linux can theoretically be made almost instantaneously. The web is more like the San tribe with its loosely-defined nucleus and lack of hierarchy. IMPACTS usually prefer that kind of structure.

An SN structure, after the creative-formative-productive agents have enabled its creation, becomes the status quo and therefore prone to being opposed to change. The strength of this resistance will depend on how strong the nucleus is, and how much IMPACTS energy is infused throughout the system, or how easily it can flow into the structure. This resistance to change seems to be a universal model though the structure is not always successful at resisting. A hurricane is a good example. A hurricane moves westward against prevailing weather patterns, bringing badly needed moisture to areas generally baked by the summer sun. It has to force its way into the established structure. The same thing is attempted in political revolutions. Both the hurricane and the revolution are examples of powerful IMPACTS change energy coming from the periphery, moving against prevailing structural forces.

It appears that the key to maintaining structural balance is to keep the doors open to the original creative-formative-productive IMPACTS energy on which the structure was founded. That energy generally serves as a reliable, positive change agent. Structures get into trouble when they deviate significantly from their foundational energy.

The non-inclusive energy flow and overall rigidity of the hierarchal structure since the development of agriculture have been problematic for many IMPACTS and their attempts to 'fit in'. IMPACTS were bred in a tight, loving circle but have to live in a dispersal-oriented hierarchy—and supply that hierarchy with the energy that is essential for its functioning. It sounds like a prisoner situation, doesn't it? For many IMPACTS, it is. Sometimes, despite their most sincere and concerted efforts, living in the hierarchal structure of society will just not work for IMPACTS, which often leads them to blame themselves for not being able to adapt. IMPACTS need to understand that their profile developed when times were different, when peaceful, cooperative living was the only way. The last 6,000-10,000 years have turned everything upside down while the core of the IMPACTS profile has remained the same.

In the next chapter, we will explore how the dynamic of IMPACTS bonding applies to our contemporary world.

Summary

When the SN structure started assuming control of the creative-formative-productive IMPACTS energy, the biggest difficulty for the SN was how to 'tame' the IMPACTS and incorporate them into the new order, because without them, there would be no new order. Of course no one had any idea of the underlying dynamics involved and still don't, for the most part. Taming IMPACTS would prove to be a not-so-easy task and continues to have its difficulties today. The hierarchal structure and the circular, change-agent IMPACTS still do not have a perfect marriage. It is two totally different forms of energy, just like the two different energies in the atom—the pulling-inward nucleus and the creative-connecting electrons.

The story of civilization is the story of that marriage and how the various 'nuclei' (SN leaders) use their 'valence electrons' (IMPACTS) against the IMPACTS of other SNs as they compete for power and control. The alphas use their captured energy against the captured energy of other alphas. The challenge for the SN alphas is how to get more innovative and productive IMPACTS within their domain. It is not that simple because IMPACTS are always on the lookout for better opportunities. If things become intolerable, they will vote with their feet—if they are not hindered from doing so.

As agriculture advanced, everything would change except the core of the IMPACTS, the San-like descendants of the San tribe and the San-shaman. Human society would be built around the IMPACTS, but instead of being independent, they would now be in a largely subservient role to the SN. When possible, the IMPACTS have driven civilization forward, sometimes in cooperation with the SN, and sometimes in opposition to it.

The SN tries to keep the IMPACTS away from politics on the national level because their transparent approach to problem-solving is not welcomed. The SN prefers that the IMPACTS gravitate to areas that enhance the power and control of the SN. The economic arena is favored because in today's world business is the source of power for the SN.

The SN is accreting the energy of the IMPACTS just as a black hole accretes matter and energy for its survival. Both the SN and the black hole must accrete because they do not produce their own energy.

The new hierarchal structure moved quickly to change the WE consciousness that originated with the San to more of a ME consciousness, emphasizing the personal responsibility of the individual for his own well-being and minimizing the connections among individuals. The nuclear family assumed the major survival responsibilities that had been shared early on by the entire San tribe.

With the SN in control, consensus and democracy were out and directives were in. The SN was not part of the feedback loop. In the San tribe, everyone had been part of that loop.

Chapter 8

Today's World - Accretion and Production

Today's world little resembles the world that the San began developing over 100,000 years ago. Their world had no boundaries, very few possessions, little if any politics, and rare discord. The San lived in small groups where egalitarianism was the norm. Men were caring and nonaggressive—women were strong and independent. Both men and women were androgynous in almost every respect. That is what was needed in order to survive the environment at that time. A San tribe was a mobile group that revolved around the creative-formative-productive shaman. Creative-formative-productive energy (CFPE and IMPACTS energy) and the ability to be mobile seem to reside together, and generally this type of energy is peaceful—unless provoked.

After 100,000 years or so of the San and their descendants laying the foundation for modern human society around the world, people discovered the benefits of agriculture. Consequently, over the next few thousand years, the development of a significant SN (structural-nucleus) began to take shape. The roots of some early SNs likely formed in more landlocked areas, such as the steppes of present-day Ukraine, Russia, and Kazakhstan, north of the Caspian and Black Seas. It is believed that the horse was domesticated in this area 4,500 to 5,500 years ago.

Like the nucleus of an atom, constricted mobility such as would be found in landlocked areas could possibly mean a different kind of energy—more of an accretive-dispersal energy. CFPE is on-the-move energy, as is the electron and as were the San. That is why today you see IMPACTS being very mobile, and why throughout history, artisans and artists have hopped from society to society, wherever their skills were needed. Artisans, artists, and traders have always been like honey bees, spreading culture and ideas near and far.

The modern world is all about the continuation of what started with agriculture and really with the first atom—accretion and production. It is the same principle everywhere you look. Everything is a 'binary star' where one is accreting from the productive other. Marriage is a binary star—humanity is a binary star—a business is a binary star. The most important issue though, given that all of us accrete to some degree, is: What is the nature of the relationship between the accretor and the producer? Is it benign? Is it loving? Is it fair? Is it exploitive? Is it brutal? Are both being enhanced, or are both being damaged?

The actual story of recorded history is not what we have been taught. Yes, there have been struggles between and among cultures and states and continue to be, but it did not start out that way. The first war that occurred after the development of agriculture was between the hierarchal, mostly patriarchal SN, the accretors (the herdsman ants), and the circular, community-oriented IMPACTS forces, the producers (the mealybugs). An excerpt from Jared Diamond's *Guns, Germs, and Steel* provides an example of what has occurred thousands of times since the SN emerged.

"On the Chatham Islands, 500 miles east of New Zealand, centuries of independence came to a brutal end for the Moriori people in December 1835. On November 19 of that year, a ship carrying 500 Maori armed with guns, clubs, and axes arrived, followed on December 5 by a shipload of 400 more Maori. Groups of Maori began to walk through Moriori settlements, announcing that the Moriori were now their slaves, and killing those who objected. An organized resistance by the Moriori could still then have defeated the Maori, who were outnumbered two to one. However, the Moriori had a tradition of resolving disputes peacefully. They decided in a council meeting not to fight back but to offer peace, friendship, and a division of resources."

The offer was never delivered, but it would not have mattered. The Maori killed hundreds over the next few days, even deciding to cook and eat some of the bodies. They killed most of the remainder over the next few years, as it suited them.

The Moriori were a small, isolated group of peaceful hunter-gatherers, very much like the San. The Maori were farmers from a dense population on New Zealand's North Island who engaged in constant warfare. Both peoples had diverged from the same Polynesian group less than a thousand years before. One group had settled down, enabling an SN to form. The other group was still moving.

This is the ugly story of history that has not been told. It happened all over the globe in thousands of different locations, and it is still happening today, though the victims are rarely hunter-gatherers. Sometimes there has been a less violent assimilation of groups, but unfortunately this was often the outcome—death to the entire peaceful IMPACTS group. And when they died, their peaceful, innovative genes died with them.

The balance of forces rapidly started tilting away from the San-like IMPACTS and toward the SN wherever agriculture became the lifestyle. With agriculture, the IMPACTS could be captured and their energy used to perform work and to build structures, including weaponry and ships, which could then be utilized to accrete from other groups. Remember how cyanobacteria started the whole food chain with photosynthesis? It is the same with the IMPACTS. They are the beginning of the human civilization 'food chain', and it started with the San and San-shaman.

The Maori killed the Moriori because the Moriori objected to being slaves. In other words, they objected to having their creative-formative-productive energy accreted, and thus being imprisoned. They refused to be mealybugs. But if the Maori couldn't utilize the energy of the Moriori, then from the Maori viewpoint, they were just in the way and should therefore be dispersed through destruction.

A brutal, extreme SN tries to capture all the CFPE within its domain, just like a black hole. That is why there is so much agony and suffering around war—people are being dispersed in horrible numbers and in horrible ways. It is the opposite of IMPACTS energy, which is seeking ways to prevent dispersal and to tighten the bonds among people. Do you recall how we mentioned that antimatter is found around black holes, and around binary stars where one is accreting from the other? War is the same—it is antimatter for human life, and it too is found in proximity to accretive SNs.

The IMPACTS have a tough job because the predominant energy in the universe is move-apart, or dispersal, energy. As we all know, it is much easier to destroy something than it is to build it. Brutal SN-types destroy and walk away, leaving the producers, the IMPACTS, to fix the mess. Do you think the SN comes in after a mistaken bombing run and buries the dead, cleans away the debris, starts the rebuilding process, and issues apologies? Not a chance. They leave all of that for the IMPACTS and their supporters.

Our world is not much different from that of the Maori and Moriori. The same dynamics are operating, but they have been dressed up so we can't see the true spectacle. In our recent past, slavery, colonialism, and the subservience demanded from women were different aspects of the same thing. It was all about accreting—capturing—energy and resources that would aid in extending the reach of the SN-domain. Slavery increased workable energy; colonialism added markets and increased access to resources; women could have more children, providing more soldiers, laborers, businesspeople, engineers, and females. The results were being funneled to the top.

As Jared Diamond makes clear, *guns, germs, and steel* have destroyed huge swaths of humanity, including, I might add, millions upon millions of IMPACTS innovators. The SN has been naturally selecting its own favored, productive mealybugs and discarding the rest. The SN needs natural resources, and it needs IMPACTS to develop them, if it is to have a well-functioning machine. The countries and empires throughout history that have attained that level have had access to both.

The twentieth century provides an illustration of the vivid contrasts in the battle between the two forces, the SN accretors and the IMPACTS producers. A cursory glance will show that the army of the SN can summon far more members to its side than can the army of the IMPACTS, and that SN army can do horrendous damage. In World War I, about 20 million people were killed, and in World War II about 70 million were killed.

It is always amazing to realize that very few people have any desire to fight and kill in the first place, yet the whole world went to war on two occasions in the 20th century, not to mention all the wars *on the side*. How does this happen? Why can't the IMPACTS side mobilize war resistance that is just as powerful as the SN war machine? It is because of the nature of the interaction between the two energies. Accretive energy is stronger since it is part of dispersal energy.

We have to keep in mind that the goal of the SN is power and control, and to accomplish this it has to break the cohesion of IMPACTS energy and its original bonding and orientation, redirecting it in ways that enhance the SN. The SN has been extremely successful in that quest, but that is no surprise because that is how dispersal energy behaves universally. For verification, all we have to do is look at Hubble photos of the universe, and we see billions of similarly-designed and constructed galaxies. The nucleated structure appears to be a hybrid of the two energies with the dispersal force (nucleus) exerting more power and control.

You can see why we have the world we do. The peaceful genes, such as those of the Moriori, do not stand a chance against passionate SN desires to capture what it wants and/or believes it is entitled to have. The only reason that the IMPACTS are still around is because they are indispensable to the functioning of this civilization. If all IMPACTS like the Moriori had been killed, the fuel and light needed to run the species would have been extinguished.

The story for the IMPACTS since the emergence of agriculture has been how to live with and address the accretive-dispersal tendencies of the SN, no matter the time or place. Like the nucleus of an atom, it is a small core, but it is powerful and casts a very big shadow. As many of us learned in world history, the Romans 'conquered' the Greeks, but the Greeks were actually the civilizing influence on the Romans. The Greeks supplied abundant IMPACTS energy, along with the Etruscans, whose engineering skills the Romans accreted early on. The city of Rome was actually built in Etruscan territory.

The Romans captured the creative-formative-productive right brain of the Greeks and Etruscans in order to provide the fuel— the IMPACTS energy—for their own masculine left brain. That is the story of civilization in a nutshell. The mostly-peaceful IMPACTS are providing the innovation and the civilizing influences at the same time. But as I alluded to earlier, if IMPACTS are provoked, or if they experience severe injustice, they can be as nasty as anyone.

In countries, creative-formative-productive IMPACTS energy and structural-nucleus (SN) energy rarely grow together in an equal fashion. A cursory glance at history tells the tale. The SN grows in a direction away from IMPACTS principles and then overreaches, resulting in extreme pressure on what is left of cohesive IMPACTS energy. The SN attempt to break the bonds of other countries often backfires, leading instead to the dissolution of its own bonds.

Once the SN oversteps, it is very difficult for it to reclaim its original position on the totem pole. The graveyard of historical powers provides some examples: the Greeks of Pericles and of Alexander the Great, the Roman Empire, the Mongols, Britain during its colonial period, Germany, Japan, the Soviet Union, and others. They have all had their day in the sun, and though most have recovered, they have not assumed their previous lofty status. But that lofty status was built on the currency of power in use since agriculture, and that is military strength.

Military power is tempting for the SN to utilize, but it is obviously a double-edged sword. It can easily create more problems than it solves. IMPACTS enable the typical overreach because the SN can accrete their CFPE, but the IMPACTS also enable recovery. When the militaristic SN head has been chopped off, a solid foundation of IMPACTS usually remains to put things back together, if it is possible. They certainly have a better chance of doing so with a new SN. IMPACTS societies, by and large, did not and do not try to take over other IMPACTS societies or SN-controlled societies. Societal accretion is predominantly SN behavior.

The expansionist actions of the Romans as they gobbled up one society after another depict the SN philosophy of accretion in practice. Eventually, they bit off more than they could chew, and everything fell apart. The IMPACTS who had been captured could not save them because the continual expansion kept stretching and fragmenting IMPACTS bonds. The producers could not keep up with the accretors, or they did not want to.

The first war after the establishment of agriculture, exemplified by the massacre of the Moriori, was of course won by the SN. You can see from the above example that the San-like people (IMPACTS) had no idea how to deal with such a brutal, uncivilized force. Though large numbers of IMPACTS were wiped out in this early stage of dispersal, enough were captured to supply sufficient innovative energy to keep the SNs going and developing. The SNs could then proceed to the next war—against each other. It would be my-captured-IMPACTS against your-captured-IMPACTS, my mealybugs against your mealybugs.

That war is clearly still ongoing, but the first war is not over either. Around the world, insurgency groups are fighting against oppressive SN actions. Again, the power-control SN uses its IMPACTS to defeat the generally more 'just' goals of other IMPACTS. Part of the tragedy is that IMPACTS almost always get along well with other IMPACTS anywhere in the world—if they are left alone. But the SN will not leave them alone if it needs them or if they get in the way.

Of course, not all insurgencies are IMPACTS-led, and not all SNs are snuffing out IMPACTS. But it is clear that much of it is going on and has gone on in the past. The story of the universe is the attempt to strike a balance. Human civilization is not exempt from that struggle though the SN people would have us believe that their actions are honorable, and that those who are objecting are terrorists. We must remember that the SN is often a fearful entity with little conscience, and that its primary goal is power and control. (Think again of a black hole.) Therefore, we can expect to hear almost anything from the SN—and we do. And it sounds very similar to what the opposing SN is saying.

Myanmar

Those IMPACTS who are closer to the SN, the SN-IMPACTS, are the foundation of the economy, anywhere in the world. A modern economy cannot function without these IMPACTS. It is not possible. But other elements of a society may depend on the P-IMPACTS, such as a non-profit for battered women, a research facility, an art museum, a library, or a bookstore. If you look around the world and see communities or countries or areas that are not functioning, it is because there is a shortage of IMPACTS, or the SN has quashed the natural energy of the IMPACTS who are there, or war may be underway.

Let me tell you a story that will reveal different aspects of the modern world and accentuate some of the things we have been talking about from the past also. The dynamics of the two major forces in human society, the structural-nucleus (SN) and the innovative IMPACTS energy, have been the same for thousands of years, and so have the human travails associated with that struggle. The following is a true story, but I have changed the names because I do not have permission to use them.

Myanmar, formerly Burma, has become a modern-day example of the extremes to which the SN will go in order to maintain control. It is believed that at least a million Myanmarese are hiding in neighboring Thailand in order to escape the excesses of the government. There is an ethnic minority in Myanmar called the Karens who make up about 7% of the population of 50 million. The Karens have been battling the government and seeking independence for decades. In 1995, Si Ko Hla, a general in the rebel army, and his family, were forced to leave Myanmar for Thailand when the government army attacked and laid waste to the area. Thus began a ten-year struggle for some kind of normalcy for the family, and for a new home.

In 2002, the family was granted UN refugee status, and after another three years, arrived in North Carolina, thanks to the help of Lutheran Family Services and a local Lutheran Church. More than 15 members of the family have been resettled in the Raleigh area. The ten years spent in limbo since leaving Burma and arriving in the U.S. were filled with danger and heartache, beginning in Thailand where the Thais are not especially welcoming of the Myanmarese.

The mother of the family actually epitomized the IMPACTS profile: she was a nurse for the rebel army. The father spoke English and made sure his children did also. Several of the six children speak many languages. The father stayed in Burma to fight with the rebel army but died in 2001 at the age of 78.

This is a story that has been repeated thousands of times throughout history—a minority IMPACTS-type group standing up to the SN, sometimes staying to fight, other times being forced to flee. It has been a major factor in the movement of people and IMPACTS genes around the world, and one reason female genes appear to have had a wider dissemination than male genes. Just imagine what a journey in search of freedom would have been like at any time in the past 6,000 years without the UN or Lutherans to help.

There are other IMPACTS elements in the story in addition to the IMPACTS versus SN struggle: positions of critical responsibility in the rebel army (nurse and general), survival skills, the push by the parents for the children to be educated, the ability to live in different cultures, a positive attitude, determination and perseverance, innovation, and something we see often in regard to the IMPACTS—new beginnings.

Chinese Artist

In modern China, IMPACTS economic energy has been unleashed in the past 30 years or so. But that was to be expected once the Chinese leadership decided on a more market-oriented economy. China has such a rich history of San-like civilizations that its IMPACTS roots run very deep. Plus, in the past, it has been a crossroads for many different peoples, which is always a sign of extensive IMPACTS activity and influence. Where you find melting pots and cultural intermingling, you find copious numbers of IMPACTS. The U.S. is another example. Again, IMPACTS are generally borderless. IMPACTS cover the earth as electrons cover the nucleus. And like electrons, IMPACTS are light and mobile.

All is not well in China. The environment is suffering terribly, and China's past attention to healthcare for the masses is being deemphasized, unless one happens to have money, which sounds a lot like the U.S. attitude. Economic, political, and military power are growing, but again, those who are accumulating the power are not the IMPACTS, but rather those who comprise the structural-nucleus. It is the accretion-production process again. The IMPACTS produce and the SN accretes. The Chinese SN wants IMPACTS energy working in the economy, but it does not want it operating in other fields such as human rights or politics. But the U.S. SN is the same; it too wants IMPACTS energy mainly confined to the economy. Around the world, it is the same dynamic and has been since the birth of the SN.

Let's take a look at a Chinese artist, Zhang Yimou, to see how the SN energy field pulls the IMPACTS into its realm (accretes it), and how it has done so repeatedly for the last 6,000 years or so. You will note that the IMPACTS energy field does not pull in the SN—the SN pulls in the IMPACTS. As noted earlier, the

IMPACTS can get along quite well without the SN, but the SN is lost without the IMPACTS. The IMPACTS attract the SN because of their (IMPACTS) productive abilities, but they do not accrete the SN. They produce and therefore attract, and the SN accretes.

Zhang Yimou's father fought as an officer in the Nationalist army against the Communists during the civil war that ended in victory for the Communists in 1949 under Mao Zedong (Mao Tse-tung). When the Cultural Revolution began in 1966, Zhang Yimou, at age 18, was sent to the fields to live and work with peasants. After five years, he discovered art and photography while working in a cotton mill as a machine technician, and was determined to make it his future. In 1976, he was accepted to attend China's only film school.

Zhang's early films painted a less-than-glowing picture of China. In 1994, he was prevented from attending the Cannes Film Festival where he had won an award. When he was nominated later for an Oscar, China pressed for the nomination to be withdrawn. But when the Olympic ceremonies were performed in Beijing in the summer of 2008, Zhang was the director of the glittering spectacle. He also sits on the country's top political advisory committee. It has been quite a turnaround for the former rebel.

Other Chinese artists are realizing that putting politics aside can make for a much less stressful life, and maybe even produce riches. This is the template that has been used over and over by the SN to draw the IMPACTS over to its side. "If you give up politics in your art and thinking, you will be rewarded." Each time a rebel artist moves over to the SN side, it makes it more difficult for an alternative viewpoint to be heard, and it strengthens the SN. Gradually, non-status quo voices become little more than whispers, and the SN domain becomes very homogeneous, and powerful.

Freedom of speech and expression is relative. We like to think that we are a bastion of free speech, but are we really? The general theory of relativity applies to societal thinking just as it does to Jupiter. The SN intent is to get everyone stationed in the proper orbit, serving the SN. Again, just look at a spiral galaxy. The dynamics are the same.

Barbarians and Philosophers

The true differences that exist among people today and have for the past few thousand years are due in large part to SN leaders who attempt to accumulate power and control, and thus divide human beings. All cultures have as their foundation the IMPACTS and IMPACTS energy, which is basically a duplicate of the original energy of the San tribes and their shamans. You will recall that for tens of thousands of years, modern humanity was mostly composed of San tribes, all basically the same, or duplicates of one another. The difference today is that there are now SNs built on top of this San and shaman (IMPACTS) foundation, and they are all different.

After the culture has started its development, it can go in many different directions because the creative-formative-productive IMPACTS energy will, in most cases, become subservient to the SN energy. It will be captured and embedded. Therefore, it will lose much of its influence in the future direction of the culture.

An example is the founding of this country. There was plenty of IMPACTS energy among the founding fathers, but as the structure grew (the country and its SN government), IMPACTS energy dissipated, as it usually does in most structures, spreading out to fill other roles such as those in the economy. IMPACTS energy was transformed from its 'new beginnings' role to its innovative-maintenance-improvement-change role. It became largely subservient. Consequently, it was not long before the country took on a very less-than-excellent SN orientation. Gone were the likes of Thomas Jefferson, James Madison, and George Washington, and entering were presidents like Millard Fillmore, Franklin Pierce, and James Buchanan.

The Greeks called those who did not speak the Greek language barbarians, and that included Romans, Persians, Phoenicians, and Egyptians, all very advanced societies. But everyone else did basically the same thing. It is an SN trick that has been around since the emergence of agriculture and chiefdoms. The IMPACTS in those societies most likely did not refer to other people as barbarians, just as IMPACTS today do not. Today, it is still the SN and their supporters who call people such names. Now, the most popular term is terrorist where a few years ago it was guerrilla. But the landscape has changed from jungle to desert and other desolate terrain. The names change but the dynamic remains the same.

Plato, Socrates, and others of the time believed that only 'philosophers' should be leaders. But philosophers at that time actually meant philosopher-scientists—those who were continually searching for knowledge and trying to improve the plight of humanity, or basically IMPACTS people. Today, it is very difficult for philosopher-scientists to break through the SN-installed barriers.

The absence of women in politics and the impediments to their involvement are clear evidence that the structural-nucleus wants few IMPACTS attitudes in its sphere. Remember, IMPACTS energy is mostly female-oriented, and includes a 'new-beginnings' element. An SN wants stability, not new beginnings. The attitudes toward women and the participation of women are usually clear signs of the extent of IMPACTS energy within a structure. The San had egalitarian attitudes about gender; today's IMPACTS are usually the same. If you see something different, you can be certain that SN energy is prevalent.

According to the Congressional Research Service (CRS), about 17% of the members of the 111th Congress are female. Other minorities include: African-Americans—about 8% of the total; Hispanic-Americans—about 6% of the total; Asians and Pacific Islanders—about 2% of the total. There is one Native-American in Congress. White males, though they are only 36% of the American population, dominate the halls of power. But as we have noted, male energy is mostly accretive, so there should be no surprises here.

Also in this session of Congress, according to the CRS, there are 57 lawyers in the Senate and 168 in the House of Representatives—about 42% of the total membership. From my research and according to how I define IMPACTS, I found that only about 1% of IMPACTS are attorneys. These are some other occupations represented in the 111th Congress: 16 medical doctors, 2 dentists, 3 nurses, 2 veterinarians, 1 psychologist, 1 optometrist, 1 clinical dietician, 1 pharmacist, 4 ministers, 3 physicists, 1 chemist, 6 engineers including a biomedical engineer, 1 microbiologist, 5 accountants, 2 professional musicians and 1 semi-professional musician, 1 screenwriter, 1 documentary film maker, 3 organic farmers, 1 auctioneer, and others. These occupations, about 10% of the total, are suggestive of IMPACTS thinking. But how much of it is captured SN-IMPACTS energy and how much is more independent? It is hard to see an IMPACTS influence on the finished product—legislation and edifice philosophy. We cannot forget, however, that this is an SN-IMPACTS hybrid world, and the IMPACTS are largely serving the SN.

Human civilization, past and present, follows the same 2nd Law of Thermodynamics that the physical world follows—concentrated energy disperses when not hindered from doing so. IMPACTS hold the concentrated energy together that the SN uses to sustain itself. But this IMPACTS energy also works for justice and equality, which can cause problems for the SN. Currently, however, the SN has the IMPACTS energy well under control.

Political economy

Today's world combines the traditional, horrific battlefield with the economic battlefield. Whoever controls the flow of goods, services, and information has the upper hand. Any threat to that control is a serious problem for SN leaders. If you are not strong economically, you will not be strong militarily—unless you have a sugar-daddy country that provides protection and other assistance. Today, it is all about economic growth, and alliances of course, which are necessary for trading and military purposes.

The U.S. is the current world economic power, although the whole world is mired in an economic slump, precipitated principally, most would agree, by U.S. policies. Other countries were seeing robust economic growth before the crisis struck; e.g., China, Brazil, Russia, and India. When a solid recovery gets underway, it would not be surprising to see a strong push for a new international financial arrangement because many countries are dissatisfied with U.S. leadership. Because the dollar is the principal reserve currency in the world, or the currency used for major transactions globally such as oil and gold, the U.S. has certain advantages, such as not having to worry so much about budget deficits. Many countries are concerned about the future value of their dollar reserves.

To be an economic power, you have to have copious numbers of IMPACTS within your domain, and you also have to have access to markets and resources around the globe, including millions of other IMPACTS. A strong military is required as well in order to protect 'your interests', wherever they happen to be. The structure keeps getting bigger and bigger with more and more needs that have to be satisfied, which means more IMPACTS energy that must be continually infused into the system. It is all about connecting the nodes of IMPACTS energy and innovation. The more connections, the more potential energy and power. The SN is drawing on that concentrated IMPACTS energy just as a hummingbird draws the nectar from flowers or a backyard feeder.

Countries in the world today are a dualism of politics and economics; hence the term political economy. Here again, it is a binary situation. The political part is accreting from the economic part. The business sector is captured just as the mealybugs are captured by the herdsman ants. The concept of globalization, by spreading manufacturing, customer relations, and other business functions around the globe, has the potential to complicate the political side of the political economy. Domestically, however, it is a well-defined symbiotic relationship. The SN accretes the benefits from the economic side and basically does what it wants—within limits. There are safeguards in most countries, but the SN can usually circumvent these if it so desires. The Americans and British did it recently in Iraq as they did everything possible to convince the world that Saddam Hussein was an imminent threat. If an SN wants to do something and has the IMPACTS strength to draw upon, it will do it and make up an explanation later.

What is the business side getting out of this symbiotic relationship? The same things that the mealybugs are getting from the ants—protection and the opportunity to create, produce, and enjoy a 'good life'.

Countries and corporations are following the same general hierarchal model and working in tandem, which can be a powerful partnership—for good and for bad. Businesses can be very beneficial, and we also know that they can be unscrupulous and detrimental to human beings and the environment. It seems the bigger they get, the easier it is for them to abuse their situation— very similar to an SN government in that regard. Brute force takes over as the foundational IMPACTS energy gets pushed aside.

Business is number one around the world and in the U.S. because it has proven itself to be the best way to keep most of the natives fed and the 'least-restless', and to enrich the SN at the same time. With each delivery, the SN is strengthened. Therefore, business gets special treatment. You will notice who got all or almost all of the bailout money after the financial meltdown ensued—BIG business, not the besieged homeowner or small business. The government arguably caused the primary problems by allowing the protective bonds to be weakened—and then left those adversely affected to pick up their own pieces. This is a typical SN attitude. "We do what we want and you, the individual, have absolutely no say-so in it. We also accept no responsibility for the consequences of our actions. We are not your partner. We are in charge." The SN is not part of the feedback loop.

SN priorities are obvious. Yes, doing nothing during this crisis was not an option; swift action was needed to avert further disaster. But wouldn't it have been better to develop a program that aided big businesses, small businesses, and individuals? Instead, it was the usual trickle-down concept—help the guys at the top and some of it will trickle down to the rest of us.

Communist countries like China, now existing mostly in name only, have learned that they too have to have an engine from which they can accrete. Therefore, they have adopted the capitalist model. If energy is to move to the top, everyone seems to have learned that it works best when both parts of the machine, the political and the economic, have the same basic design. Of course, many people get left behind under the capitalist model, but that is a 'peripheral' concern for the SN, which is why many of its 'problems' often emanate from the periphery.

In today's world, just as it was in the ancient past with Greece, Persia, Egypt, Rome, and others, the societies with the richest blend of IMPACTS, along with access to natural resources, generally produce the most expansionist and potentially aggressive attitudes and behaviors. These attitudes of expansion and aggression can be in the political-economic realm or the military realm, or both, and will exist on a continuum. What we see on the world stage are predominantly male SNs utilizing the tangible and intangible benefits provided by their own IMPACTS, along with benefits from associations with other SNs and their IMPACTS. The ultimate goal of course is to exert power, control, and influence. So, now as then, those who can summon the most IMPACTS energy and resources generally rule the roost.

The world is very much like the wolf pack configuration we mentioned earlier. Nobody wants to get the omega label, but the various alphas vying to be top dog will not hesitate to propagandize their choice or choices. In the wolf pack, the position of omega appears to be a way for the pack to unload excess aggression. Is it any different among nations? Aren't countries doing the same as they depict their adversaries in a dehumanizing fashion? It wasn't that long ago that the Vietnamese were characterized by the U.S. SN as a people who did not value human life, except those Vietnamese who were on our side. Now, Vietnam is almost an ally. Of course, being an ally has nothing to do with valuing human life. Is war as much about men unloading aggressive tendencies as anything else?

Most countries are like the bulk of the wolf pack—they are constantly vying for the most favorable spot they can get. With wolves, the alpha can be dethroned, and the pack must be ready to shift alliances. You see the same behavior on the world scene with people and countries.

People ask why we continue to have an endless cycle of wars and other such depravity. The answer is simple—the SN people control the world, and they do not think or feel or view the world as do the IMPACTS. Wars are extreme male energy, but females can start and participate in them also. The SN has set up the world according to its own liking. It has the microphone and the keys to the military vault, and it is adept at obfuscation. It is creating a reality, but it is unaware of the specific dynamics involved. It thinks it can continue bending and shaping the structure of reality—that it can sculpt it. That is where the IMPACTS, especially those on the periphery, come in—they try to maintain balance and keep the SN from going too far. The SN, as expected, generally tries to make these peripheral IMPACTS out to be disloyal troublemakers. By understanding the dynamics at work and the power that rests with the IMPACTS, we may be able to alter the reality that the SN is crafting, though it will take time and effort.

An example of this kind of thinking was provided by an official in the Bush administration to Pulitzer Prize-winning journalist and author, Ron Suskind, in 2002. According to the

Bush aide, the media and others are in the "reality-based community"—those who "believe that solutions emerge from judicious study of discernible reality . . . We're an empire now, and when we act, we create our own reality. And while you're studying that reality . . . we'll act again, creating other new realities, which you can study too . . . We're history's actors . . . and you, all of you, will be left to just study what we do."[1] Another official said, "We're not big on nuance."

That is the story of the past 6,000 years or so. I cannot explain it any better than that. That is from the horse's mouth. That is how they think—and behave. The IMPACTS always have much to do after such utterances, a lot of repair work, because destruction is almost always involved.

If we have a human profile that fits the valence electron, which is the most extreme electron energy, in that it is farthest from the nucleus and most apt to semi-escape from its pull, then we most likely have its opposite. This opposite would be a black hole-type profile that mirrors the extremes of the nucleus, the power-control center. And a cursory look at history and the contemporary world makes the case strongly. So we have both extremes casting their energy on the world, one possessing innovation and anti-status quo attitudes, often loosely engaged with the structure, and the other more like a black hole, trying to suck everything into its domain. Most of human life operates in the middle, but it is hugely affected by both extremes.

Eliminate Potential

We are living in the world that the SN has created over the past 6,000 years or so, and we are seeing it the way the SN wants us to see it. Our version of reality is coming from the 'herdsman ants' (the SN), not from those who are actually providing the energy, the mealybugs (IMPACTS). And as noted before, the human SN is not nearly as aware of the dynamics involved as are the ants. The ants can see it because the mealybugs are a different species. For us, it is all wrapped up within one species, and political correctness (PC) does not allow us to see it. Let's cut through the PC and the omnipresent propaganda and try to get a better perspective on what is truly happening. Keep in mind that what we are hearing on a day-to-day basis is extremely crafted, and only loosely based on the real story. Again, it is like pre-Copernicus Europe—what the Church preached bore little resemblance to the true dynamics. It is no different today.

In the world of physics, the 2nd Law of Thermodynamics is trying to eliminate potential. Potential is defined as a difference in concentrations of energy. So the 2nd Law wants to disperse everything and destroy the potential—to create homogeneity. You will recall that in the scenario proposed by many astronomers and astrophysicists, a sea of black holes will form before the eventual dissolution of matter. So that is where we are in the development of human civilization at this particular moment—the black hole SNs have formed though there are varying degrees of 'blackness'.

I said early on that if we want to truly understand how the world works we will have to take ourselves out of our own 'sculpted' reality and put ourselves back into the natural world. The question though is this: Do we really want to understand the underlying dynamics or would we prefer to continue with our game of 'denial and compromise of truth'? I suspect it is the latter for most of humanity because of what we have learned—the 'mass' of the SN allows it to control and direct much of societal belief just as the nucleus controls electron energy that is closest to it.

Warring and adversarial countries want to do what the 2nd Law tries to do—they want to destroy the potential of another country, or disperse its concentrated IMPACTS energy. They want to disrupt IMPACTS nodes and sever the connections among these nodes. Israel has been threatening to attempt to do this to Iran militarily as Iran works on its nuclear program. Iran says its work is for peaceful purposes but of course no one believes that. Israel is certain a nuclear bomb is the goal, and a nuclear bomb in the hands of the Iranians would provide too much 'potential' to suit the Israelis. Therefore, they want to destroy the program and disperse the threat. Currently, Israel is the only country in the Middle East with nuclear weapons, and Israelis want to keep it that way. However, Israel has never admitted publicly that it has nuclear weapons and is one of the few countries which have not signed the nuclear non-proliferation treaty, though others who possess nuclear weapons have signed the treaty, including the U.S.

Let's look at the Middle East and see how it fits into our model. The Middle East today is a metaphor for what has been happening the past 6,000 years. It has everything we have been discussing: an SN attempt to create a new reality, accretion of land and resources, 'my IMPACTS versus your IMPACTS', loads of SN propaganda with very little relationship to reality, massacres and murder in the guise of self-defense, people who are defending themselves being labeled 'terrorists', and aggressors with no interest in cause and effect.

God has also been interjected into the fray as many Jews and Israelis believe that God promised them the land of Israel and that they are his chosen people. My attitude is that people can believe whatever they want as long as they are not hurting others. When beliefs are destructive of others is where I draw the line. I will not sit by silently and watch others being abused, exploited, dehumanized, and murdered. Such behavior is the antithesis of IMPACTS energy, of which life itself is the ultimate manifestation. If we sit and watch, the power-control forces grow ever more menacing.

In today's world, political correctness demands that writers and others give a 'balanced' account in their writings and opinions—if they want to be accepted into the 'respected' corridors of the mainstream community. But this works to the benefit of the SN as it keeps people from exploring the genesis and dynamics of the issues. And it enables the status quo to remain firmly in place.

The same balancing forces that we see operating in the atmosphere, manifested as weather, are operating within the human realm as well. The 2nd Law of Thermodynamics is always 'on' no matter the specific location in the universe. It is all about balance and concentrations of energy. The human SN is trying to attain power and control, and outside of its awareness its energy is directed toward the capture of IMPACTS energy, or its dispersal if it is a threat. Human IMPACTS energy 'thinks' in terms of justice, equality, opportunity for all, abandonment of no one, human bonding, creative-formative-production, improvement, and the protection of value. These forces are vying constantly though the SN clearly has the upper hand because it is a brute form of energy.

People have learned for the past 6,000 years or so that longevity is increased, most of the time, when you defer to the SN and its decisions. That 'attitude' was probably initiated in the universe 380,000 years after the Big Bang as the male protons captured the female electrons. The same capture scenario has continued on down the line to us. The male SN has done the same; it has power and control over much of our thinking, which is also electrical in nature. It seems to be a proton versus electron universe with the proton as the force-carrier of the male dispersal force and the electron the force-carrier of the female bonding force.

People avoid expressing their opinions about the Middle East like the plague, which is normal given the seemingly insoluble nature of the problem and the incendiary atmosphere surrounding it. But that insoluble nature is telling us clearly that objective IMPACTS are not involved to any significant degree because if they were, there would be movement towards resolution. It is also telling us that this is an SN alpha domain. In the animal hierarchy, you cannot protest the behavior of the alphas—unless you are willing to accept the risks of doing so. This situation is saying the same: "This is alpha territory and you are treading in dangerous waters."

People have learned to their chagrin that they can pay a heavy price for 'swimming in these waters'. The SN has said "Leave it to us," which flies in the face of IMPACTS expressive tendencies. These tendencies developed from the San tribe over 100,000 years ago as individual expression on every topic was encouraged, where boundaries of almost any kind were unknown, and when cause and effect was the standard of inquiry.

The script we are hearing and have been hearing from the government and the media about the Middle East conflict is absolute fiction in that the dynamics and motivations are not being identified properly. In Israel, people may express their opinions about Israeli behavior, but in the U.S., it is considered off-limits. I will try to approach the subject with a borderless mentality. Human beings might have a hard time believing that human civilization is following the same energy patterns of the atom, but I contend that everything is following the same model, and that includes the Middle East.

Since most human beings prefer to 'invent' explanations for human behavior or just accept the directives of the SN rather than dig deeply for cause and effect, many will not like what I have to say. But as anyone will be able to see, what I have to say is following precisely the model we have been exploring.

We have seen in our discussion that the IMPACTS and the SN are the two major forces in human society, and that the SN exists only because of a strong IMPACTS foundation. Everything that is happening among human beings is built around these two forces. The Middle East situation is like most everything else human—it is simple in its dynamics but appears complicated to us due to the attempts of the SN to 'create' a new reality and manipulate our perceptions of what is happening. Layer upon layer of falsehoods are piled one on top of another.

First, let's look briefly at part of the history of the Jewish people. The actual story may be different from what we have heard previously. It is believed by many scholars that the early Jewish people originated with the Haberu, a group of wanderers and 'outcasts' who traveled about in the Middle East. These people tended sheep and were agriculturalists, stonecutters, and warriors. Each clan had a strong, capable military arm that utilized mostly guerrilla tactics. Some of their early leaders, Abraham, Moses, Joshua, and David, were also very capable military men. Abraham, the father of the Israelites, said that there is one God and that this God promised him the land of Canaan for his people.

A group of these early Israelites traveled to Egypt around 1400 BCE because of drought conditions in Canaan. Some scholars believe that while in Egypt they served as mercenaries, protecting a critical northern area. However, the Pharaoh, fearing their rise in numbers and strength, made them laborers on public works projects. Still, such laborers were treated well, fed well, and provided with medical care. There is no indication that the Israelites were slaves, though it may have felt like slavery because an SN force had control over their lives.

Moses asked the Pharaoh around 1250 BCE for permission to leave Egypt with his people, and the permission was granted, even allowing the Israelites to leave with arms. But it appears that the departing Israelites attacked a town and stole supplies for their long journey, after which Pharaoh gave chase.[3]

Let's look at this with the IMPACTS concept in mind. Obviously there was an abundance of IMPACTS energy within the Israelites along with a strong protective element, which is perfectly normal. We have noted before that when you find plentiful IMPACTS energy (female-oriented), you have the potential for extreme SN formation (male-oriented), which can easily take control of the situation. It is the same in a galaxy; black holes form and take control as stars form.

As stated early on, in the post-agricultural world, those who possess both SN and IMPACTS energies can be particularly successful at attaining their goals, though their goals may not be favored in some quarters. The Jewish people, to a significant degree, appear to have copious amounts of IMPACTS energy as well as copious amounts of power-control energy. It is not a surprise that such a situation exists if you understand the IMPACTS concept. The power-control element is the flip side of the IMPACTS element. It is the 'house' and the IMPACTS element is the foundation.

The Jewish people seem to have a large percentage of SN-IMPACTS where the two forces have merged together in individuals, with the SN actually being the Jewish edifice. The SN force and the IMPACTS force, though basically opposites, can exert tremendous power and control when working together—with the SN in charge. (You can see the same occurring with a galaxy.) The degree to which we see the extremes of the two energies within individuals in a relatively small group may be unique in the world. The synergy of these two forces is why Jewish accomplishments are so impressive in some areas and so frightening in others. There is obvious 'star' behavior and obvious 'black hole' behavior.

Of course all of this exists on a continuum, and is the same dynamic we have been seeing throughout our discussion—male SN energy using female IMPACTS energy as its fuel supply. A significant percentage of the Jewish community has few power-control elements and abundant IMPACTS elements, and a significant percentage has mostly power-control elements powered by IMPACTS elements. In Israel, a core of extreme power-control people is in charge which is absolutely no different from the situation in Iran and many other countries around the world. Almost all nucleated structures have such a core but of course the power-control energy varies greatly from one structure to another. As an example, the SN of General Motors may not be able to save the company in a crisis but the SN of the country would not let the U.S. dissolve. Companies come and go but countries usually have strong staying power due to their immense power-control core.

The Israelites left Egypt intent on taking over Canaan with whatever means were required. Canaan had many flourishing areas. The Israelites had no intention of assimilation—they planned to conquer. The only crime of the Canaanites was that they resided on lands that the Israelites claimed as their own— because 'God' had given the land to them, a God that Abraham had 'discovered', or created. One God makes the whole process of accretion much easier if you claim that HE is on your side. A male God is much more likely to go along with accretion and dispersal than is a Mother Goddess, which many San-like people of the time believed was the ultimate power. A Goddess would have favored assimilation and bonding—not accretion and exclusivity.

The Israelites, in their campaign to take the land, killed everyone and everything that they encountered—men, women, children, goats, donkeys, and anything else that was breathing. They wanted the land bare so it could be populated with their people. Did the Israelites consider the Canaanites to be terrorists or savages for defending their land and people? I suspect that they did. Dynamics do not change. An accretive mentality sees only enemies—it does not see potential partners.

As we saw with the San, a small group living in genetic isolation can maintain the same basic behavior and attitudes for tens of thousands of years. The leaders of modern-day Israel exhibit strikingly similar behavior and attitudes to those of the Israelites 3,000 years ago. Today's Israelites are blaming the Holocaust for their attitudes on security and accretion, but is that really the truth or are rationalization and denial at work?

It might be worth noting that Einstein, who was Jewish, thought that the Bible was a "collection of honourable, but still primitive legends . . . For me the Jewish religion like all others is an incarnation of the most childish superstitions. And the Jewish people to whom I gladly belong and with whose mentality I have a deep affinity have no different quality for me than all other people. As far as my experience goes, they are no better than other human groups, although they are protected from the worst cancers by a lack of power. Otherwise I cannot see anything 'chosen' about them.'"[2] Today they are not protected from those cancers because now they have immense power.

It seems utterly ridiculous to post such a quote but it is also ridiculous that people would think that they are chosen by God and that God has promised them a piece of land. A sense of entitlement, needless to say, distorts the perception of reality as accretion and power-control issues predominate. Being 'chosen' by God necessarily means that your adversaries were not chosen by God, and therefore, they have less importance. This condescending attitude can obviously lead to abuses as we have seen. No people are any better or more important than any other people. Anyone who believes so possesses a delusional belief system. All cultures are IMPACTS-based; it is largely the character of the SN that distinguishes one from the other. Jews and Palestinians are actually close genetic cousins but Jews seem to want nothing to do with their Palestinian relatives. Mostly it seems they just want them to disappear.

If Israeli behavior were utilized by any other country, that country would immediately be condemned and ostracized from the international community. Their behavior is inhumane, in their treatment of Palestinians under their occupation and in their wars with neighboring countries. I presume that the present group that controls the international community, led by the U.S., feels guilt about the Holocaust and therefore gives Israel license that other countries do not receive. Plus, there are other factors involved, mostly economic and political.

Israel appears to be able to do anything it wants, and doesn't have to take responsibility for its actions. It is coddled and enabled, especially by the U.S., and therefore behaves like it is coddled and enabled. Other countries are expected to take responsibility for their actions, but not Israel. Their slate is always wiped clean after particularly savage behavior—wiped clean by the Israelis and their supporters but not of course by their victims. It appears that the philosophy of its supporters is that if Israel can be assured of security, it will eventually 'grow out of its reckless, destructive teenage behavior'. But will it, or are the issues more complicated than that?

We have seen that accretion along with dispersal is characteristic of an atomic nucleus. The electron does not accrete—it connects, shares, and bonds as it forms a new structure, a molecule. Within humanity, we have also seen that the SN accretes but not the IMPACTS, except when the IMPACTS are part of the hybrid group, the SN-IMPACTS. Furthermore, energy that originates from the SN can be violent, just as dangerous gamma rays emanate from the nucleus of an atom. Peripheral energy, like that of the bonding valence electrons, is generally peaceful—again, unless provoked.

When trying to ascertain the derivation and causes of violence around the world, look for accretion and the consequent severing of concentrated bonds, along with the attempt to reorient the flow of energy and resources. That is usually the most important sign.

That is what an SN or any nucleus is attempting to do. In the Middle East, Israel is accreting Arab lands and resources, and in the process it has broken long-established bonds in the area. This is clear SN behavior and strategy, or an effort to control the neighborhood by 'eliminating potential'.

Let's look at some examples of accretion or attempted accretion in the past: the U.S. in Vietnam, the Roman Empire, the Soviet Union's move into Afghanistan, the Soviet Union's grab of Eastern Europe after World War II, Europeans' emigration to southern Africa, Hitler's aggression in Europe, the Spaniards' massacre of the Incas. The examples are endless. What we see in all of this is a country or entity going into another area already occupied by a different race or nationality and proceeding to try to take control of it, strongly influence its behavior, or destroy it and take the resources. The accretors and the aggressors are one and the same, and they are coming from the outside. It is extreme male behavior. Usually, accretion and aggression end poorly for the perpetrators because they are going against well-established IMPACTS roots which can be very resistant and resilient.

It is the same in the Middle East. Israel is the accretor and the aggressor. The Palestinians are not attempting to accrete. They were already there. Much of their land has been stolen, and they have been subjected to unimaginable indignities and massive suffering. They are trying to protect themselves, their people, and what is left of their land and resources. The world barely voices any concern and continues to enable Israeli behavior. Why not consider sanctions for Israel until it behaves in a humane manner?

Yes, the Jews have been subjected to similar atrocities and indignities, including the Holocaust, but the Arabs had nothing to do with that scourge. History is replete with unimaginable horror and the destruction of many different peoples, including Native Americans, but that gives no one the right to use past suffering as an excuse to continue the brutality.

When people are being mistreated and worse, IMPACTS elements will rise up in an attempt to protect them. It appears to be a universal law. Think again of maternal behavior in any species—it is ferociously protective. That is IMPACTS energy. When the 'destiny-fulfilling' white settlers tried to seize lands long held and revered by Native Americans, the Native Americans did not hand them over—they fought to the death to protect their loved ones and their lands. In the process, they were tagged as savages, just as the San were in Africa by the Europeans. This is typical, extreme SN behavior. If you happen to occupy land that an extreme SN wants and claims as its own, and you fight rather than surrender, then today you are a terrorist. But make a note that from today's perspective, most of the people that the various SNs have called savages and worse over the ages were merely doing what was natural—trying to protect themselves and their people from massive slaughter. The real savages were the SN perpetrators.

I do not know anyone who does not want the Jewish people to be able to live in a safe, secure environment. But the actions of the Israelis engender the opposite of safety and security. Their inhumane behavior against people in the Middle East, including killing others with an appalling indifference, is an affront to humanity. U.S. and Israeli policies leave one with the impression that the Jewish people are the most important people on earth. While important, they are no more so than anyone else, and of course they should be held to the same standards as others. Why wouldn't they?

Would the U.S. and the white-controlled international community have enabled and permitted the Israelis to carry out their killing and theft of land in a country of Caucasians? What if Israel had been set up in Switzerland or France or England? Would we have the same carnage or would dead white people be unacceptable? Think about it for a moment. In the Introduction, I mentioned that the most pernicious assumptions are the ones we don't know we are making because they appear so intrinsically obvious. Some of us are assuming race is not an element in this conflict, but I think it clearly is. We deny a lot of unpleasant reality so we can be comfortable.

The goal of the SN is to get people on its side who behave as it wants them to behave. It wants passionate partners but it will settle for silent ones. The SN edifice wants followers who believe as it does—that it is OK for the SN and its friends to commit atrocities and murder but not for those designated as enemies (omegas). The hang-up is that usually the IMPACTS have a justice problem—they support it—and that can lead to difficulties for the SN and the IMPACTS. But the U.S. SN has been wildly successful at muzzling the people on this issue.

The policies of the U.S. and Israel have driven a wedge between people in the Middle East and around the world, but again, dispersal is part of the makeup of SN forces, and the creation of enemies is part of dispersal. Today, we see terrorists everywhere, but who is producing these terrorists? Were they born to be suicide bombers or did they develop into them? What is a terrorist anyway? Were the inhabitants of the American colonies terrorists for opposing the powerful British SN, or were they freedom-fighters? Were the American Indians terrorists for defending their lands and people, or were the settlers the terrorists? It all depends on whether you are looking at it from the SN side or the IMPACTS realm. The SN usually sees sides—the IMPACTS generally see a bigger picture with human bonding elements.

Israel appears to feel no responsibility to try to work with and get along with its neighbors, which to me is the first responsibility of any country. Imagine if an Arab country or Iran imprisoned Jews in occupied territories and forced them to live in subhuman conditions, and then unleashed one of the world's most powerful militaries on innocent civilians, killing thousands. What if an Arab country or Iran blanketed civilian populated areas with cluster bombs as the Israelis did in their 2006 war in Lebanon? Forty percent or so of the bombs did not explode, becoming land mines for farmers and townspeople. Ninety percent of the cluster bombs were dropped in the final 72 hours of the war when it was apparent that a ceasefire was imminent. Unfortunately, this behavior appears to be normal for Israel rather than aberrational.

This is the kind of behavior that must be counterbalanced by IMPACTS around the world if the current human species is to survive. The extreme SN preaches the golden rule to its citizens while it murders non-citizens. Such SNs are depending on those of us who see the obvious hypocrisy to remain quiet and passive as they go about their deadly tasks.

As we noted, SNs neither learn from their mistakes nor see them as mistakes. Even though it should have been obvious that everything about the Vietnam War was a horrible blunder, that the Vietnamese were driven more by nationalism than by Communism, and that China and Russia were not part of a giant monolithic Communist machine but rather had competing interests, none of that mattered to the power-control SN-types in the U.S. The insane destruction continued unabated for many years. The craziest thing about the Middle East may be the effort to create something that is so unjust, and then to call those who disagree terrorists.

Just as the valence electron energy gets relegated to the periphery, so too do the opponents of SN behavior, principally the IMPACTS. It is very hard to change extreme SN behavior, from either the inside or the outside. It is an energy that is generally not amenable to change, whereas IMPACTS energy is the opposite. The war in Iraq, started by the U.S., is another example of the SN effort to impose a reality. As is usually the case, the SN is befuddled that everyone does not fall in line and 'get with the program'. But as I said, cause and effect is not the strong suit of the SN, or really even an interest.

How did the U.S. happen to become the principal benefactor of Israel and its supporters in their drive to create a 'new reality' in the Middle East? Largely because our political system is run by money and special interests, and segments of the Jewish community and supporters of Israel have taken full advantage. They have consistently positioned themselves at strategic points in the SN power and defense structure, as did Richard Perle, Paul Wolfowitz, Douglas Feith, and others during the run-up to the Iraq invasion. Friends of Israel had been clamoring for U.S. initiated regime change in Iraq for many years. The events of 9/11 provided an opening, and Israel's supporters rushed in. In the past couple of years, they have been pushing for similar action in Iran.

Many of these 'public' servants seem to have an agenda in mind rather than service to the country of the U.S. The agenda appears to be to make Israel strong and secure no matter the costs to the U.S., and as mentioned, those costs have been extremely high—in blood and money. It is interesting how different the attitudes of the two countries are toward assimilation. The U.S. is a true melting pot and Israel doesn't even want Palestinians to drive on many of its highways, ostensibly because of 'terror' threats, but realistically because Israelis want total segregation from everyone except Jews.

Through powerful lobbies such as AIPAC, Israel's supporters have kept a stranglehold on debate and stoked the flames of divisiveness. IMPACTS try to create solutions that will aid humankind; extreme SNs try to create the reality they want and to which they feel entitled. This is fierce energy that will do just about anything in order to accomplish its aims. As I said, extreme SN behavior can be a natural consequence of copious IMPACTS energy because there is so much surplus production and innovative-productive potential. It is there waiting for someone to use it—or abuse it. An extreme SN is all too happy to take control.

In recent years, American foreign policy has effectively become American-Israeli foreign policy, with the U.S. actually subservient to Israel's desires. The Israeli SN has actually captured the American SN, as strange as that may sound. But it is not strange when you understand the IMPACTS concept and its dynamics. The links between the two countries are bound in complex ways, including in the governmental realm where much of the linkage is 'behind the scenes'. Israel and its supporters have adroitly learned how to accrete vast sums of financial aid, power, and weaponry from the U.S. Spying on their only real 'ally' has also been frequently utilized.

At its base, religion is a major factor in the relationship as right-wing Christians typically are big supporters of Israel and its policies, forming a powerful alliance that thwarts legitimate and needed inquiry. The public is blocked from the discussion, which takes place only behind thick walls. It is similar to taxation without representation—we are providing the taxes but our opinions are not being allowed representation. Our anger at being used to enable such destruction is also not getting through the very thick walls.

I have never heard the word justice used by either the Israelis or the Americans when they discuss prospects for what they call 'peace'. Peace in this context seems to mean acquiescence, subservience, and advantage. Why not just drop the whole charade and tell the truth? "We are not interested in justice for the Arabs." And then maybe add: "The concept of justice plays no part in our foreign policy." Then everyone would know what some of us already know. Justice was a primary consideration of the San but it has little attraction for an SN because it interferes with goals and intentions. If somehow you could create advantage out of justice, then the SN would enthusiastically favor it.

IMPACTS, unfortunately, are enabling the situation in the Middle East by having their energy used and abused by the two SNs, the governments of Israel and the U.S. Some IMPACTS are fighting against the injustices while others are turning their heads to the carnage, knowing it is being done in their name and with their money. IMPACTS around the world have such a visceral reaction to Israeli behavior in the Middle East because it is the opposite of their prescription for how humans should be treated. It violates every line of their code of ethics. The world often finds anti-Semitism on the rise after particularly brutish behavior by the Israelis. I suspect it is usually not the Jews who are hated but rather insanely destructive and inhumane policies. Whatever happened to the concept of cause and effect? It has largely disappeared because it has very little value for the SN.

Where you see this kind of alpha behavior, you necessarily have to have those who will assume their lowly position unconditionally, or they will bear the consequences of not doing so. Usually alphas pick on other races, and it is no different in this case. Arabs and Muslims have become the perennial omegas as propaganda has turned the world against them, even as they continue to supply the world with much of its petroleum needs. But we must remember—SNs divide people because that enables them to accomplish their power-control goals with more efficiency. Race is a convenient way to do that.

We have fallen a long way since the days of the San when right was right and wrong was wrong—when people searched for answers in order to make everything better for one another and for all. The San thought peace was normal; we think war is normal. We have to understand the derivation of this change in identification of what is normal and what is not. It is coming from the SN faction. As long as IMPACTS have little influence in the power corridors, we will have these problems. But I want you to see it for what it is—it is not human nature or any other such rationalization. It is a lack of IMPACTS involvement in decision-making—a lack of political power.

The U.S. has sacrificed tremendous amounts of goodwill, money, international standing, and human blood as it has enabled the Israelis to continually snub their noses at basic standards of decency and humaneness. Rather than using its tremendous supply of IMPACTS energy to provide positive contributions around the world, the U.S. has allowed the Israelis and their supporters to divert its attention and resources from other needed areas and concerns. It is as if the world revolves around the desires of the Israelis, and unfortunately, to a large extent, it does. Even the Iraq war, which will ultimately cost over a trillion dollars and hundreds of thousands of lives before it is over, was, in my opinion, fought mainly for the Israelis in the interminable quest to make them 'secure', a quest that may not be possible because of their actions.

Newton's third law of motion applies: "To every action there is always an equal and opposite reaction." It may be slower in the human world than it is in the physical world but the model and the dynamics are exactly the same. In the human world it can take quite a period of time for the IMPACTS forces to reassert themselves, but unless there is total extermination, the IMPACTS dynamic is always working to restore balance. As I have stated, a hierarchal SN abhors a naturally occurring dynamic unless it is working in its favor, and the IMPACTS dynamic is usually working at cross-purposes to the SN. Hence, the SN moves to control it or destroy any semblance of it, as we saw with the massacre of the San-like Moriori by the SN-like Maori.

The Israelis now see Iran and its pursuit of nuclear technology as an 'existential threat'. I am certain that the countries in that part of the world see Israel as an existential threat, and they have every right to do so based upon its past actions. If I lived in the Middle East, I would be far more terrified of Israel than of anyone else.

It is easy to distinguish between IMPACTS behavior and extreme SN behavior. The former tries to defend itself against an encroaching force; the latter takes in (accretes) all that it can within its 'proclaimed' domain. When you look at the American settlers fighting the Indians, it should be easy to distinguish the defenders from the aggressors. It should be equally clear in the Middle East. Just look for accretion.

The SN tries to obfuscate by convincing us that we have to stand as ONE against 'all those bad guys' who are also acting as ONE. But in reality it is not ONE at all, but rather two different forces, the SN and the IMPACTS. The SN is corralling its IMPACTS forces to be used against the IMPACTS forces of others who just happen to see things quite differently.

The so-called terrorist group, Hezbollah, did not exist until Israel invaded Lebanon in 1982 and subsequently stationed its troops in southern Lebanon. Hamas was created in 1987 and received early backing from Israel as Israel tried to splinter the Palestinians, who at the time were led by Yasser Arafat. Hamas, in addition to having political and military wings, also provides social and educational services to Palestinians as does Hezbollah to the Lebanese.

This is the way the extreme SN world works. It constantly creates enemies with its policies and then casts all blame on them. This kind of behavior and reasoning has been going on since the SN came on the scene over 6,000 years ago, and reveals a tenacity that will continue to be difficult to control. It seems to always be present at different locations around the globe.

Someone might ask, "Aren't organizations such as Hamas and Hezbollah exhibiting extreme SN behavior also? Surely they are not as innocent as you make them sound. After all, they have killed a lot of Israelis and others, too."

Israel is radicalizing the neighborhood and has been for a very long time because it is behaving in a radical manner. Abnormal behavior begets abnormal behavior. What else would anyone expect—that abnormal behavior would produce normal behavior, that theft of land, murder, and constant humiliation would produce acquiescence and goodwill?

Hamas and Hezbollah started out as IMPACTS movements to protect and care for their people. An IMPACTS organization, if the energy is sufficient, will develop its own SN. It happens everywhere. It is a natural process. The same occurred here in the United States as IMPACTS energy gave way to an SN structure. It happened among the Jewish people as abundant IMPACTS energy enabled a power-control SN. Since the development of agriculture, the dynamic has been ongoing within human civilization.

Israeli actions initiate the process of 'protective' IMPACTS organizational development by the victims, and then feed it with further destructive behavior. In short order, you have a full-fledged 'enemy' SN fueled by IMPACTS energy with no way to predict the path it will take. The longer the conflict continues, the more enemy SNs will be created. Nothing less than what has happened in the Middle East should have been expected. The whole terrible saga has been totally natural in its development, with IMPACTS forces of protection vying with accretive SN power-control forces.

You will recall that the size of Jupiter prevents the formation of any other spherical entity within its gravitational domain. The asteroids between it and Mars have no chance of forming into anything cohesive. Actually, the asteroid belt is thinned continually as the rocks are thrust out of their orbits by the mass of Jupiter. Israel is trying to do the same. It wants to be the Jupiter of the neighborhood and make cohesion impossible. Up until this point, it has been very successful. But that is no surprise because the U.S. has supplied it with the most advanced weaponry in the world and an abundance of other resources. What if the Palestinians had been provided with the same resources, including modern tools of propaganda?

I have mentioned that almost any battle in any field of human endeavor for the last few thousand years has really been 'my IMPACTS versus your IMPACTS'. The Middle East is no different. The Israelis can call upon vast amounts of IMPACTS innovation from the Jewish people, and from the U.S. It is really quite extraordinary. The Palestinians, while similarly gifted, do not have vast resources at their disposal. They have to get by on sheer wit and courage. They do not have a limitless well from which to draw as do the Israelis.

This is a good question to ask to ascertain the true intent of countries: If their enemies were removed tomorrow from the scene, what would they do? Would their behavior be better, worse, or about the same? I think with Israel it is clear—their behavior would be worse. They would steal as much land and resources as possible with no apparent conscience. They are behaving just as the early Israelites did over 3,000 years ago—combining accretion with human destruction.

Such comments will immediately bring charges of anti-Semitism. But that is a tactic to intimidate and smear, no different from typical SN behavior utilized over thousands of years. It is the same as calling someone a savage; it is an attempt to dehumanize. The intent is to stop the conversation before it begins so that those with power-control aims can carry out their work with no hindrances and no questions. It is alpha behavior and thinking. A large part of the world is anti-American, but that is no reason for us not to examine our own behavior to see if we are in fact responsible for some of these attitudes. Obviously, we are. The Israelis are also responsible for much of the hostility that comes their way. But they seem immune to self-questioning. It is always someone else's fault.

Recently, an interesting idea emerged from the Israel-Palestine Center for Research and Information, suggesting that real, meaningful change once again may have to come from the periphery. The center recommended that the international community grant Palestine full membership in the UN. At that point, Israel would be occupying a UN member state in violation of the UN charter, and therefore would be forced to withdraw.

Recently, the European Union's foreign policy chief, Javier Solana, offered a similar proposal. Of course the Israelis rejected it immediately as you would expect an extreme SN to do. Power and control is their language, and that includes all aspects of the discussion.

Why am I stressing this particular area so intensely, and singling out Israel? The main reason is that I am an IMPACTS-person who possesses the attitudes I have been writing about in this book: I hate manipulation, obfuscation, and hypocrisy; I want to leave no one behind; I believe that someone has to speak for the innocent victims in these types of tragedies; I hate to see lies being accepted as truth; and I believe in the sanctity of human life. The SN often has no conscience—it destroys and it walks away, leaving the victims to somehow pick up the pieces and move on. We need to understand the 'quantum mechanics' of these types of conflicts so we can do our part to try to strengthen the IMPACTS dynamic and prevent the horrible destruction that we see on a recurring basis.

The public has been brainwashed on this subject as propaganda has been utilized incessantly for decades, adroitly combined with the volatile issue of past suffering (Holocaust). Israel is trying to do in the Middle East what the U.S. tried in Vietnam—to create a 'new reality' on the ground. The U.S. effort in South Vietnam failed because the U.S. did not understand the dynamics of the region or the true reality of the situation. The effort was doomed from the beginning. It was ignorance and insanity unleashed together with incalculable destruction as the result. As one might expect, there is also ferocious resistance to this attempted 'creation' in the Middle East.

At many of the volatile points in the world, adults are not in charge. Many of the so-called leaders are nothing more than playground bullies with horrific weaponry and state-of-the-art propaganda machines at their disposal, and no clue in the world about cause and effect or an interest in it. This, sadly, is what passes for normal in today's world with justice and other such humane principles cast to the periphery. But it appears to be the same battle we see throughout the universe—the 'unfeeling' nucleus and dispersal versus IMPACTS elements which have to bond in order to prevent dispersal.

President Obama, an obvious adult, shows signs of trying to put some distance between the U.S. and Israel. But what kind of effect will that have? The Israelis may try to machinate a crisis that will force the U.S. to 'choose' between its friendship with Israel and its present path of distancing. Anything is possible from those who are intent on accretion. Eventually, it all comes apart just as the sun finally set on the British Empire. Accretion is usually a step in the direction of dispersal. Look at the old USSR and its many attempts at accretion. Bonding is the only thing that has a chance of 'permanence' in the world but sadly that is not what SNs do as a general rule. Look at all the lives and resources wasted as SNs choose to accrete and create enemies rather than reaching out and helping other people develop themselves and their resources.

Within Israel and the Jewish community, the same SN versus IMPACTS struggle is taking place that has always occurred within societies. But clearly the IMPACTS have been no match for the SN elements and their power-control core. Still, there are strong voices of dissent in the Jewish community and in Israel to Israeli policies.

The situation in the Middle East shows us that what we see on the surface may hide deeper truths. The critical dynamic is actually within the IMPACTS part of the equation. IMPACTS have to understand that they are the key to the resolution of these types of seemingly insoluble conflicts since they are the ones who are essentially pitted against one another. What makes it so difficult is that on the side of the SN aggressor, IMPACTS energy and innovation are embedded within the entire war machine. On the rebel side, the IMPACTS are easier to identify because they are operating with few resources. Hence, they are more vulnerable. This is why it is so difficult for a rebel group to stand its ground. It is trying to survive against a heavily-embedded, IMPACTS-powered SN machine.

The Middle East is an example of the world we have had for the past 6,000 years where totally upside-down thinking passes for normalcy, and logical, justice-oriented thinking is labeled as abnormal and extreme, even dangerous. It is actually quite astounding. But it shows the power of any SN, political or otherwise, to influence the energy within its domain, and to dominate.

People have asked me why I decided to write about Israel—did I really think it would do any good? IMPACTS around the world have to start speaking out, in effect forming a powerful counterweight to SN forces and their deadly, opportunistic schemes. IMPACTS have to see themselves as a totally different force apart from the SN. Bacteria that became mitochondria retained their own DNA even as they were accreted by the nucleus to provide the energy for the cell. And even though the nucleus has gradually accreted some of the mitochondrial DNA, still the nucleus would collapse in a second without the mitochondria. That is also the case with the SN and the IMPACTS. The IMPACTS are increasingly being pulled in toward the SN but 'they still have most of their own DNA'. IMPACTS need to understand the reality of the situation and the power they possess to shape events.

The SNs want us to continue to see the world as us against them. They want us to see dispersal rather than bonding. But that is to be expected—that is the nature of the energy. The nature of

IMPACTS energy is bonding, creating-forming-producing, helping, and improving. Currently, the SN is using the energy of the IMPACTS to serve its own needs and desires. IMPACTS need to start the long process of reorientation of their energy.

Iran

Iran is an excellent example of the IMPACTS versus SN process. Iran has a deep, rich history that goes back over two thousand years. You cannot have such a culture without a strong IMPACTS foundation, just as you cannot have an atom without an electron. Like many other countries, Iran has found its resources accreted by Western powers, particularly the U.S. and Britain. In 1953, the CIA overthrew the democratically-elected government of Iran led by the popular Prime Minister, Mohammed Mosaddeq, and installed a monarchy led by the Shah, Mohammed Reza Pahlevi. This was done primarily to allow Western oil companies to accrete Iranian oil reserves. The arrangement worked well—for the West—until the Shah was overthrown in the 1979 Iranian Revolution that brought the Islamic Republic into being. Here again, we have the accretive SNs of the West refusing to take responsibility for bringing about the Islamic Revolution and the dangers that the West now faces. Cause and effect and taking responsibility are anathema to most SNs.

In such revolutions, there is always extensive IMPACTS energy involved or else it could not happen. The U.S. revolutionary war against the British was the same. But just as in the U.S. where the IMPACTS-laden founding fathers gave way to a developing SN, the same occurred in Iran, as happens with any revolution that succeeds. The more hardcore SN elements quickly took control in Iran, just as was the case in Israel. But that is no surprise because often it is the most extreme leaders who enable victory and survival.

The Iranian SN is resistant to change internally, but so are almost all SNs, some of course more resistant than others. There are many enforcers available to the Iranian SN: the Revolutionary Guards, including the Quds which operate mostly in other countries, the regular military, and a volunteer group of zealots called the Basij. The Basij motorcycle about with clubs and guns when 'trouble' arises, attempting to disperse the concentrated, rebellious IMPACTS energy. The IMPACTS crowd has no such resources on its side. That is why IMPACTS have to be stealthy and clever if they decide to join the fray. Thirty years ago, in 1979, many of those in power today were the IMPACTS rebels, and now they are part of the SN. That is the process—the SN emerges from the IMPACTS.

In the recent voting protests in Iran, women were the ones 'on the ground' who were out front. This ferocity is similar to that seen across the animal kingdom when females believe their offspring are in danger. To me, this is a clear sign of deep disenchantment. Freedom of expression is a core IMPACTS value, and it can be suppressed for only so long. Of course, freedom of expression exists on a continuum. The trick for the SN is to find the point where there is just enough to keep the natives reasonably happy but not enough to topple the power-control core.

The hypocrisy of the West is obvious in its approach to the Iranians. Western countries, including the U.S., strongly criticized the Iranian regime because of its suspect voting practices in recent elections and condemned the violence associated with protests against those practices—as they had every right and obligation to do. But barely a whimper was heard from the West during the recent massacre by the Israelis in Gaza where they slaughtered about 1400 people and wounded another 5,500. The Israelis timed their killing spree between the U.S. Presidential election and the Presidential inauguration. That way, there was really no one to voice any protest from the American side. Just imagine what would have happened if the Iranians had done something similar.

Many organizations and people are calling for war crimes trials for the Israelis because of their actions in Gaza, including the bombings of schools, hospitals, ambulances, and 'safe havens'. But this is standard procedure for the Israelis who also routinely target UN forces and facilities though it was the UN that passed the resolutions creating Israel. Of course the Israelis deny any deliberate transgressions. One has to wonder if they really believe what they say or if they just don't care about the lives of non-Israelis. Elie Wiesel, Nobel Prize winner and Holocaust survivor, says that ". . . to remain silent and indifferent is the greatest sin of all . . ." Weisel says it is the moral responsibility of all people to fight hatred, racism, and genocide. But his attitudes do not seem to apply to Israel's treatment of the Palestinians and other Arabs. The hypocrisy is extraordinary, and so is the world's acceptance of it. But sadly that is the kind of world we have today and have had for the past several thousand years.

Now the West is trying to stop the Iranians from developing nuclear weapons by demanding access to all of their nuclear programs. But where is the concern and demand for access to Israeli nuclear programs? Why are they not scrutinized as well? Are they considered to be responsible nuclear partners? If so, upon what is that based? Why isn't the West as concerned about the inhumane and appalling conditions that exist for the Palestinians as they are about conditions in Iran?

It is all a big con game—our team against their team. We can have some nasty players on our team but since they are on our team, we will look the other way. But if the other team has nasty players, we will point it out loud and clear.

The Iranian leadership is a tough, theocratic, unyielding bunch but so are the Israelis. That is the way the world works—one extreme SN begets another. The Iranian leaders may be a bad crowd but they are not accreting lands that belong to others and they are not imprisoning millions of others under occupation. I'm not defending the Iranian leaders—they are truly undemocratic, against rights for women and gays and non-Muslims, autocratic, murderous, and who knows what else. But if they disappeared tomorrow and a new enlightened group of leaders took over in Iran, the Israelis would continue doing exactly what they are doing. And if anyone started to protest, then that group would be dehumanized.

Some of the future leaders of Israel were true murderers and terrorists against the British, who controlled Palestine from 1917-1948, and of course against the resident Palestinians. Two of their leading terrorists attained the office of prime minister of Israel: Menachem Begin of the Irgun group and Yitzhak Shamir of the Stern Gang. Ariel Sharon, as Israeli Defense Minister, was found personally culpable by an Israeli commission of ignoring the potential consequences of bloodshed when, during the Israeli invasion of Lebanon in 1982, Christian Phalangists were allowed to enter the Sabra and Shatila refugee camps where subsequently over 1,000 Palestinian and Lebanese civilians were massacred. The Israeli military had control of the entrances to the camps. Sharon reluctantly resigned but was later elected prime minister. Sharon has been notorious for his hatred of and incitement of Arabs.

The forces of Irgun and the Stern Gang massacred over 100 people at the Arab village of Deir Yassin in April 1948, even though the villagers had signed a nonaggression pact with a nearby Jewish village. This and many other atrocities led terrorized Arabs to flee from their towns and villages, and added 'fuel to the fire' of Arab governments which decided to confront Israel militarily weeks later in May 1948 after Israel declared independence. In September 1948, members of the Stern Gang led by Yitzhak Shamir assassinated the highly respected Swedish diplomat Folke Bernadotte, who was the UN mediator for the Arab-Israeli dispute. Folke Bernadotte had negotiated the release of thousands of Jewish prisoners from German concentration camps in World War II. The Jewish terrorists who killed Bernadotte did not want negotiation or compromise. This element is still very much alive within the Israeli power structure, and is as strong as ever—perhaps stronger—as it has been fed and enabled for decades, principally by American largesse.

The country of Israel was formed against a backdrop of Jewish and Zionist terrorism. It appears to be the same fervor from 3,000 years ago. The extremely strong, SN power-control core of Israel and the Jewish community is heavily influencing the Jewish people and everything around them. Again, it is the general theory of relativity at work. It operates the same in the human world as it does in the physical world.

People who try to segregate themselves from others in today's world generally set themselves up for trouble, especially when they do it in such an unjust and violent manner. The IMPACTS attitude is assimilation—the extreme SN attitude is exclusivity and accretion. It is the two major universal forces at work.

Why can't people see the obvious hypocrisy in the storyline that we hear from the SN and realize that it is a constant game of SN manipulation? Because that is the way that energy fields work—most energy gets captured and becomes homogeneous while some escapes to the periphery where it is less affected and has a better view of what is happening.

Yes, the story from the Middle East is just that—a story. You might say the same about much of the everyday world, and you would be correct. But the Middle East, in my opinion, is the most vivid example of the phenomenon. Whenever you see this kind of SN violence and fierce resistance to it, you can be certain that an SN is working hard to 'craft' a solution that goes against long-established, natural movements. Think again of the American settlers moving westward or the Europeans immigrating into southern Africa. But that is what a nucleus tries to do—it tries to break up, or disperse, the cohesive, natural energy that is in place, and then take control of it. It is truly universal. It happened in the first eukaryotic cell with the nucleus and in the first galaxy with the black hole.

It is a dangerous world. There are rabid people out there who are intent on doing great destruction to others and property. But a lot of the wind could be taken out of their sails if the Israelis would behave in a more civilized fashion. Currently the Israelis are sucking the U.S. once again into their corner vis-à-vis Iran, and expecting the U.S. to do their bidding, which the U.S. appears willing and anxious to do. There are many ways to potentially engage the Iranians but the U.S. will probably take the 'stranglehold' route because that is what the Israelis and their supporters prefer.

Another note on the Israeli-Iranian-West confrontation: The Israelis may try other means in an effort to stop the Iranian nuclear program rather than military bombing. One possibility could be some kind of cyber warfare where the Israelis use their considerable skills to try to destroy the program from the inside. Don't be surprised at anything that happens but do expect tons of propaganda and mistruths.

The United States

What is the future of the U.S.? Is its influence on the wane? If so, why?

It certainly appears to be, for multiple reasons. First, the rest of the world is ascending, led by other IMPACTS-laden countries, such as China, India, Brazil, and Russia. It wasn't that long ago that most emigrants had their sights set on the U.S. Now they can go anywhere, or even stay in their own country because of the interconnectedness of the world and the opportunities that it brings. Plus, many have no doubt become disenchanted with the U.S. and its 'anti-terror' policies post 9/11.

IMPACTS immigrants are what enabled the U.S. to reach its lofty status. As those immigrants go elsewhere, there will be a discernible difference in the U.S. in different areas of human endeavor though the 'leaders' will have absolutely no idea in the world what is causing it. They will engage in the usual political rhetoric, blaming the other party and its policies.

Another element was mentioned earlier—the support of Israel. The U.S. has become increasingly obsessed with Israel over the past few decades to the detriment of its relations with many other countries, and to the detriment of what it truly could have accomplished in needed areas around the globe. But Israel and its supporters have been extremely skillful at creating and fostering that obsession, and siphoning as much available energy and resources as possible. There is no reason that the U.S. should not have good relations with Arab countries and the Muslim world, but Israel and its supporters have worked hard to limit constructive U.S. engagement with the Muslim world and the 'potential' that might develop from it.

It is extremely difficult to significantly change a structure after it has assumed its primary shape. Be on the lookout for Israel's supporters to try to manipulate Obama just as they did with George W. Bush and Iraq. No matter what the Israelis say about their 'sincere desire for peace', remember that actions speak louder than words.

One of the problems for the U.S. is that our political system is set up to generate conflict and separation—dispersal—rather than compromise and cohesion. But we know how it works—a hierarchal structure divides, and there is no more pure example of that than American politics. The SN core is firmly entrenched and money is running the show, but the innovative, people-oriented IMPACTS are in short supply. Barack Obama has gotten through, but he is up against a powerful, solidly-entrenched SN structure that will exert a strong influence over his thinking, whether he realizes it or not. He will in most cases be swept along by the tide of events and the structures that surround them and him.

Today the U.S. is run by two political parties though there is not one word in the Constitution about them. Some people carelessly say that this country is a democracy, but actually it is a tightly-SN-controlled republic. Strict measures are in place to make certain that people like me and many of you reading this book are not allowed in, and not even permitted to ask questions at a news conference. The little guys and gals have been shut out because a hierarchal structure such as the U.S. government has a distinct fear of creative-formative-productive IMPACTS energy— unless it is controlled and directed to the 'proper' places. The best way to control it is to keep it in the economic sphere, or cast it to the periphery.

That is why the world has so many insoluble problems—the structural-nucleus sets up the energy field so that the solutions-oriented IMPACTS cannot get in. The SN wants to keep IMPACTS busy creating and producing products and services, thus enhancing the power of the SN. That is what the leaders did when the pyramids and other structures were built. They isolated the artisans and innovative-skill workers, treated them well, and let them create and produce. It is no different today except that the SN does not know who the IMPACTS are. What the SN does do, however, is reward heavily those who follow its template for success. If you do, you have a reasonable chance of getting some of the 'goodies'. If you don't, you are on your own—on the periphery. But you will not be alone. The periphery is the natural home for IMPACTS.

Summary

Human society and everything else, including the rest of the universe, is part of the accretion-production paradigm. The San and their San-like descendants were the producers, and their production eventually enabled the formation of the SN, or structural-nucleus. The SN, in order to survive, has to accrete from the producers, the IMPACTS.

The first war was fought by the SN to capture the 'right' IMPACTS and destroy the 'wrong' ones. It is still being waged in parts of the world. The second war was between and among SNs, each with its own captured IMPACTS. That one is still ongoing as well. Sometimes the two different kinds of war get entangled. The SN focus of destruction (without awareness) is generally the IMPACTS and their innovation and production, including the means of that production.

Economics runs the world—it is the fuel, and all countries want as much of it as they can get. A strong economy means a better chance of being militarily secure. Countries have learned that the hierarchal SN model produces 'good' results when utilized for the economy and the government.

Adversarial countries and those at war are following the 2nd Law of Thermodynamics and trying to eliminate potential, which in physics means different concentrations of energy. Countries want to disperse potential if they see it as a threat.

Footnotes

[1] Ron Suskind, "Without a Doubt", NY Times, October 17, 2004
[2] ZME Science, May 14, 2008
[3] Bible Battles, History Channel DVD, 2005

Conclusion

We have been bred not to see a dynamic in the physical universe or among human beings, and not to even look for one. But that is because our consciousness is mostly in service to the SN-controlled paradigm, or certainly hugely affected by it. Consequently, we see what the SN wants us to see—that our struggles on earth are not related to struggles in the universe, such as matter versus antimatter, black holes eating stars, galaxies devouring galaxies, and protons capturing electrons. Wouldn't it make sense that human competition and human violence had their beginnings in the heavens, and the same with love and sharing? Since we have male and female forces on earth, isn't it likely that we have the same forces beyond earth?

All of that is outside of our current consideration because we have taken ourselves out of the cosmic fray and given ourselves license to write and produce our own play. In the process, we have infused the actors with our chosen traits and have rejected any variables that might expose our analytical weaknesses. But we have done this in response to the dispersal forces of the SN which have curtailed human inquiry, preventing us from seeing and experiencing natural connections.

How much are our scientific pursuits affected by the big shadow of the SN? Personally, I believe a lot. The SN is the Jupiter in the human neighborhood. Inquisitive energy, which is science energy, is now captured energy. During the time of the San, there were no Jupiters; unimpeded inquiry had free run. But at that time it was for the well-being of the group. Much of today's inquiry is for advantage of one over the other, which is typical SN philosophy. Therefore, it is not objective—it is an agenda.

The idea that the San are the progenitors of almost all modern humans has not been accepted by people around the world or probably even seriously considered. There can be only one reason for that—the SN forces have been extremely successful at directing human consciousness away from obvious, underlying connections among human beings. What other obvious realities are we not seeing? How truly distorted is our interpretation of what is going on in the world?

They say that the universe is what we say it is. They could add that human civilization is what we say it is, and reality too. But who are *we*? WE are those who are benefiting from the current definition, just as the Catholic Church benefited from its definition in the Middle Ages. It continued to benefit until reality could be twisted no longer. Then, with Copernicus, Galileo, Newton, and others, a new reality was revealed, a reality based more on empirical study than on coercion. But the Church is still with us, and I do not mean the Catholic Church. I mean a hierarchal entity, the SN, that draws its power and existence from the work of others, and then uses that energy to coerce others to follow its dictates. If you do not create and produce, you must control, or follow. Accretors are usually in no position to compromise because they are not producing energy. Therefore, they must take.

If you leave it to someone else to tell you what is going on anywhere in the universe, they will be happy to do so, but their answers may be more self-serving and fictionalized than anything else. The current paradigm is set up to answer needs, but whose needs? What needs are being missed, and whose needs are they?

Advanced ideas in human relations and human society do not get the same reception as do advanced ideas in technology because we do not really want to look at ourselves. So we go through the motions and come up with all kinds of theories that have no chance of explaining anything. They do, however, prevent us from learning about the dynamics of human civilization, and the SN powers-that-be are thankful for that because it allows them to continue to run the world as they choose.

Today's structural-nucleus leaders welcome the innovation of SN-IMPACTS like Steven Jobs with open arms because it means more power and prestige for them, not to mention that there might be military applications, a major consideration of the SN since its emergence. But if Steven Jobs or Bill Gates or Warren Buffet or Oprah or other IMPACTS like them have recommendations on a different process that could be utilized to choose political leaders, one which would ensure more equality and integrity in the system, those recommendations would generally not be welcomed by the SN leaders. They want people to 'know their place'.

They also want anti-status quo attitudes to remain inside the economic arena, focused on producing more and better products and services. And the SN does not want to hear any comments about the setup of the economy itself with its fairness, ethical, and environmental issues. The emphasis is on creative-formative-production because that means power, control, and enhancement for the SN edifice. The herdsman ants just want the mealybugs to produce food, which translates into energy. The SN has the same attitude toward the IMPACTS.

Why would we think that our current view of reality is correct when all the other ones since the development of the SN after agriculture have been so badly skewed? The Greeks and Romans had gods all over the place—not spirits like the San and their IMPACTS descendants, but gods. The Catholic Church said that the earth was the center of the universe until the truth could not be held back any longer. Many even thought the world was flat until Columbus reached America. It was not that long ago that slavery was an accepted means of 'capturing energy', even by the so-called civilized. Around the world, women are still trying to attain the rights that San women have taken for granted for over 100,000 years.

The universe, including human civilization, is all about the 2nd Law of Thermodynamics and the 2nd Law-breakers. Were it not for this law-breaking, bonding energy, we would certainly not be here. It appears that it is the interplay between these two forces that produces a nucleated structure. Most everything we see in the universe seems to be based on this model in which the accretive nucleus appears to have more control over the structure than does the energy-producing part. After all, if something is producing continually, it doesn't make sense that it would have the ability or need to accrete. We can see this principle at work in a binary star system where the dying star is accreting from the productive one. So a nucleated structure is going to have dispersal tendencies of the 2nd Law and anti-dispersal bonding elements, or what I call IMPACTS energy.

Early modern humans like the San accreted little from the environment. They were like cyanobacteria, which were the first photosynthesizers. The San took the energy that the environment offered and innovatively used it to produce their own. Today's human society largely depends on the energy-production and bonding of the IMPACTS, the profile that developed from the San and shaman. Almost every human structure has to have this IMPACTS productive-bonding energy or it will not last.

In today's world, computers can store that energy, and people can call upon it as needed, just as Upper Paleolithic people in Europe pressed their hands on cave paintings in order to withdraw potency. Plus, by connecting computers, bonding can be increased along with available energy. That is why businesses and organizations can literally run on a computer. The computer plays the role of the San-shaman: it is solving problems, and the actual solutions can be stored, retrieved, and duplicated. Computers could be considered concentrated IMPACTS energy. That is why IMPACTS usually love them—computers can help them do so much more. And that is what IMPACTS want to do—continually duplicate good stuff, as far and as wide as possible. It is all about reaching everyone, and leaving no one behind.

IMPACTS energy appears to be the opposite of the 2nd Law. The 2nd Law is directing the universe toward a vast sea of nothingness while IMPACTS energy appears to be trying to do the opposite—to create structures of permanence. That is what Einstein did with his famous formula, $E = mc^2$. He created a permanent structure to aid in deciphering the workings of the universe. Knowledge is concentrated power with vast potential energy. Even the invention of heaven is a quest for permanence. IMPACTS attempt to do the same in the real world as they go about the business of trying to create permanent institutions that will provide a measure of equality, opportunity, justice, democracy, and protection.

We have noted that IMPACTS energy aspires to make everything the best it can be, including life itself. But as we have seen, after the IMPACTS creative-formative-productive energy starts the development of the structure, it has little power over its direction unless it remains a major part of the entity. Oprah and her enterprises are a good example of the creative-formative-productive energy remaining significantly involved with the structure. In human society and civilization, IMPACTS energy is only a small part of the total energy, but it is powerful in its ability to change the structure.

Overall, human civilization has become like most other structures as the pure, foundational IMPACTS energy of the San and shaman has been diffused, dissipated, and diluted. The part of IMPACTS energy that can be utilized by the SN is captured by the SN, and the part that has little value to the SN moves toward the periphery, where it is generally ignored.

Galaxies have their black holes and their stars galore. Countries and societies have their governmental entities that generally do not create, and then they have their IMPACTS and supporters who do. Companies are the same, an inner management core surrounded by producers. Human beings have become the same. Our brain and consciousness have become nucleated as the 'structured' left male half controls the creative, productive right female half, or at least that is the preferred model in today's world. Hence, our consciousness acts like any other nucleated structure—it focuses more on itself than it did during the time of the San when it was more of a right brain consciousness. And just as nucleated galaxies in the sky are drifting apart from one another, so too are we as individuals, and as different societies. It is the way of the universe. It is the 2nd Law in action. The original San tribe and its consciousness were concentrated, anti-dispersal, bonding energy—IMPACTS energy—that helped hold humanity together, and helped it survive.

Nucleated structures are competitive because the accretive center is male energy. We see it in galaxies as they capture and merge with other galaxies. Countries try to be top-dog and many are willing to sacrifice as much human life as required to get there. Companies vie with other companies. But none of it would happen without foundational IMPACTS bonding and energy production. What nucleated structures do is take their innovative, IMPACTS bonding energy and use it to compete against other nucleated structures and their innovative, IMPACTS bonding energy.

Important Concepts to Keep in Mind

- Structures generally begin with a creative-formative-productive IMPACTS stage. Examples: bacteria producing oxygen before more complex forms of life could emerge, the San tribe laying the foundation for modern humans, entrepreneurial activity before the business takes a more definitive form.

- The creative-formative-productive stage, if it flourishes, is generally followed by a nucleated stage. Examples: prokaryote gave way to eukaryote, San gave way to the SN form of society.

- The nucleated structure takes control of the creative-formative-productive IMPACTS elements and uses them for its energy supply. Examples: bacteria became mitochondria, cyanobacteria became photosynthesizing chloroplasts, and the San and shaman became IMPACTS who now provide the energy for human civilization.

- The nucleated structure also uses the captured creative-formative-productive IMPACTS energy as the innovative-maintenance-improvement-change energy for the structure. Think of the mother and her roles within the family. She is the creative-formative-productive element and the innovative-maintenance-improvement-change element.

- Duplicative structures such as a franchise have all the SN and IMPACTS elements embedded within the structure. It is the same for a human cell with all information needed for duplication contained within the DNA.

- Accretion is a key element in survival. All of us accrete to some degree, but it is the male SN that is almost always accreting. IMPACTS accrete very little—they share and create and form and produce. But SN-IMPACTS will usually have both accretive elements and IMPACTS elements. Bill Gates and Steven Jobs are examples.

- Change generally comes from the periphery—good change and bad change.

- The San and shaman provided the kernel from which all of humanity grew and developed, and the IMPACTS possess the modern-day personality profile of the San and the San-shaman.

- The IMPACTS are now the kernel from which the modern world grows and develops, though the IMPACTS are increasingly transferring their 'DNA' over to the SN, which is what the SN strives to achieve. Just as a star that gets knocked out of the galaxy no longer contributes anything of value to the galaxy, so it is with IMPACTS who reside on the periphery. As perceived by the SN, they are not contributing value unless they can be pulled inward. Many IMPACTS remain on the periphery and work to counterbalance SN excesses, or just ignore the SN which is also ignoring them.

- Everything in the viewable universe, including human civilization, appears to be built around the tension between the 2nd Law of Thermodynamics and the 2nd Law-breakers. The hydrogen atom, with its single proton nucleus and its one electron, is a manifestation. The proton appears to be an agent of the 2nd Law (dispersal), and the electron, when it has the opportunity, is bonding and hindering dispersal. Everything appears to be based on that model.

The structural-nucleus of humanity, the SN, was born from the creative-formative-production of IMPACTS energy. It is still on life-support; it cannot survive without the energy and production of the IMPACTS. Like some teenagers, it thinks it knows everything but it cannot live on its own. The SN depends on the IMPACTS and always will.

With little understanding of how its power was created, the SN also has little understanding of how its power can be lost. Therefore, the SN will characteristically be reckless, assuming that it can do so and suffer no consequences. Again, the SN neither understands cause and effect, nor does it aspire to understand it. Consequently, it can destroy the beneficial effects of IMPACTS no matter how many IMPACTS a society has, if it aggressively overreaches. History and the contemporary world provide plenty of evidence. But after the destruction, IMPACTS energy can help societies recover, as has also occurred many times. IMPACTS are the bacteria of humanity—they create and produce and they repair and aid recovery. The unpredictable variable is the SN; you never know what it and its supporters will do.

Finding Balance

The IMPACTS do not realize the power they have to shape events because they have never recognized their role and position in the process. As noted, the current paradigm does not allow such objectivity and deliberation. Political correctness does not permit the division of society into two distinct parts, accretors and producers, the SN and the IMPACTS. The SN-controlled paradigm wants you to see the 'atom', not its constituent parts, and specifically not society's two diametrically opposed energies.

Like the mealybugs, the IMPACTS ultimately control everything—creative-formative-production, the economy, politics, science, art, all of it—because they provide the energy and fuel for all of it. But the SN alphas have adroitly usurped the power and continue to use it for their own ends, claiming credit if things go well and blaming others if they do not. As noted, the herdsman ants seem to have a much better awareness of what is actually happening within their society than does the human SN. The mealybugs know what they are doing—their purpose is to supply food to the ants. The ants, too, know what is happening—they are totally dependent on the bugs for food. But because human society contains one species instead of two, the SN has been able to prevent awareness of the real dynamics operating underneath. When the SN used slaves, it was very obvious as to who the producers were and who the accretors were. And when the SN took over another country with another race, the dynamics were out in the open. But now the lines are blurred as the SN has become extremely adept at obfuscation and manipulation.

Where do we go from here? The alphas are firmly in control of the world structure and are playing their one-upmanship games that place us all in constant danger. Life is a game of control to them—of resources and of the power to affect events 'on the ground'. They will go as far as they can go. Extreme SNs will not stop until they hit a powerful brick wall, which in today's world means a formidable military machine. But the universe is no different. A black hole-controlled galaxy will gobble up whatever it can.

The world is following the model of the wolf pack precisely, though there are variations. The betas, the IMPACTS, are trying to reduce tensions through bonding, and are scrambling to answer unmet needs. The extreme SN alphas are generally pushing the envelope as far as possible, something that an alpha wolf would not do. But then the wolves have been around for millions of years and therefore know better. Today's world is loaded with omegas, who are usually a different skin color. Sad but tragically true. Even though we humans have almost no genetic diversity, relatively speaking, the extreme SN mentality will utilize even the tiniest differences in order to divide.

If the universe is under the control, largely, of dark-dispersal energy, then what chance do we humans have to change our current situation, given that human civilization appears to be following the exact same template? Since the universe has been at this a lot longer than we have, it has had plenty of time to strike a balance of sorts, though that balance is still tilted to the dispersal forces. But were it not for the dispersal-hindering, bonding forces, everything would be moving apart much faster. The bonding forces help delay the inevitable just as the parents' love for the children helps delay the inevitable passage into adulthood and full exposure to SN forces. IMPACTS energy helps to mitigate the harshness of a reality based ultimately on dispersal.

First, we must realize that even though this is the most perilous time in the history of human beings, it is also the most hopeful. Why? Because all of the tools are in place for connectivity, which is the foundation of IMPACTS energy. It has never been easier to communicate with anyone around the world, and to store, transfer, and duplicate knowledge. The potential is truly astounding. The reason these tools are in place is because they were invented by IMPACTS, just as the printing press was invented by Gutenberg and the worldwide web by Tim Berners-Lee. The SN usually takes the invention and utilizes it for its own goals just as it does with any form of IMPACTS energy. But just as Martin Luther utilized the printing press to spread his challenges to the Catholic Church, the IMPACTS find ways to use their inventions to further IMPACTS principles and values, one of those being human connection and bonding.

Obviously, the world needs more balance, and this means more problem-solvers in critical positions. And as we have seen throughout the book, the problem-solvers, the IMPACTS, are also the 'get along well with one another' people. Strife is big business, one of the biggest. Think what would happen to the world economy if peace were to truly break out. The threat of war is embedded in our economy, just as are the IMPACTS. Our present economy depends on both because it is SN-controlled.

Reversing this paradigm will not be easy. The world may be able to survive as presently constructed, with the SN exercising the power and the IMPACTS supplying it, but personally, I believe there are better ways for humans to live other than the script we have been using for the past 6,000 years. We need to bring things back to a working equilibrium, one that functions well for all concerned.

Is There Hope for Change?

The question today, though, is this: Is the residual influence of creative-formative-productive IMPACTS energy—San and shaman energy—strong enough to save us from the serious shortcomings of the structure of civilization that have developed over the past few thousand years?

To change the world and do it without further violence, a thoughtful strategy must be adopted. Perhaps it should be focused where the structural-nucleus derives its power—the economy. After all, that is where the IMPACTS have the most influence. That is the vulnerable spot for the SN and the power spot for the IMPACTS. If the mealybugs wanted to change their arrangement, they would be smart to utilize their control over the food supply as a bargaining chip. It is the same with the IMPACTS—if they want to change the current structure, it might be wise to look first at the economy.

But as I have said before, there are really two kinds of IMPACTS: SN-IMPACTS, who are closer to the SN and the mainstream, and P-IMPACTS, who are often politically astute but possess little clout in the economy. So the economic route will probably have only limited success. The best thing that IMPACTS can probably do to effect change is to consistently stand up for their principles. But even that is not easy in today's world as the SN works studiously to homogenize the population, ironing out all wrinkles. The SN wants a smooth path; it makes it easier to accomplish goals.

I am not advocating a violent revolution. I am advocating a quiet but open one, similar to the one led by Martin Luther King, Jr. Do you think the SN would have ever started on the path to doing the right thing were it not for Dr. King or someone like him? Eventually, maybe. As I said, the SN is not looking for solutions. It is looking for ways to maintain power and control, and the status quo—if it is in its favor. Solutions generally come from the periphery, where Dr. King resided, where the view is clearer and the tug from the SN is weaker.

After civil rights laws were passed, the Republicans initiated a 'Southern strategy' in elections whereby racial tensions could be reignited if necessary in order to corral the white vote for conservative Republican candidates. This is typical SN 'innovation'—manipulative, dehumanizing, and cynical—as opposed to IMPACTS innovation, which is generally uplifting to humanity, or attempts to be.

The world is run by a very few people. As the criminal investigators always say, "Just follow the money trail." There is a huge barrier separating regular, everyday people from those who have truly captured the controls of the world. Big business and government have combined their forces synergistically and are able to accomplish prodigious tasks, oftentimes of a diabolical nature, such as the U.S. war in Iraq. Big business provides the weaponry, and the SN puts it to work.

Big business and the SN have the same basic hierarchal structure with the same basic power arrangement—mostly men at the center. Governments are often just window dressing, smokescreens for the real powerbrokers who are as different from the average Joe as you can get. But now that we are increasingly pulling away the facade and revealing the structure of humanity as it truly is, we can start thinking about change.

What steps are needed?

First, I think a new mindset is needed. People need to realize that it is not 'us against them'. It is the SN power-control forces against the innovative, people-oriented IMPACTS forces, which also happen to be generally peaceful—unless provoked. The problem is not in some far-off country like Iran. It is not between the Palestinians and the Israelis, not the IMPACTS of those two groups, anyway. It is between the SN and the IMPACTS, the same struggle that has been around for 6,000 years.

We have to see the structure of society as it is—it is a herdsman ants and mealybugs arrangement. Almost all societies in the world have the same setup, though there are certainly varying degrees of it. For example, the Scandinavian countries and New Zealand seem to have a more balanced version of the two human forces, the SN and the IMPACTS, than do many other countries. Some would suggest that the reason is due to their peripheral locations, and I would agree because the periphery has long attracted the IMPACTS.

Essentially, international relations come down to this: "I've got my mealybugs and you have yours. If you are my friend, we can join forces and our mealybugs can do great things together, such as business expansion and trade, cultural exchange, and scientific endeavors. Plus, you can utilize the protection of my military umbrella. If you are not my friend, then I will use my mealybugs and the mealybugs of my friends against you and your mealybugs. I will try to limit your potential—or eliminate it if needed." So the SN leaders don't really have anything in the fight—they are using their mealybugs, or the IMPACTS, and other people who help supply the SN with energy. But we know that from looking at recent wars. Executive and Congressional leaders rarely send a loved one to the war zone. They send someone else's loved one. Maybe that should be a requirement: if a Congressional member votes for war, then he or she has to place a loved one in harm's way. I bet that would slow down the rush to bomb and kill. They might quickly discover communication and compromise, and respect for others.

Of course, when you confront the SN, you could be asking for trouble. The SN can be extremely dangerous, sometimes behaving like a caged wild animal. Its energy can be particularly ugly and

Conclusion

vicious. IMPACTS do not generally have energy like it, unless they are threatened. As I have said, fear resides in the nucleus because there is no creative production there. Accretion is the only way to survive.

Another part of the mindset that needs revamping concerns the basic hierarchal structure of present-day society. This structure has a clear message: those at the top of the organization know best, whatever the organization. It also says that truth flows from the top, whether the top is the President, or the governor, or the 'expert' on foreign relations, or the child therapy expert. The hierarchal society assigns value, and therefore, some slots and their occupiers are more valuable and important than others. The SN hierarchy also defines success and failure, insinuating that those who do not meet its conditions for success are thus failures. Many people take this to heart and assign the label to themselves. We need a new definition of success in this society, one not related to satisfying those conditions set by the SN. Is a person who never had a chance to finish high school because he or she had to take care of the family a failure? Is a company executive a success even though he rejects reasonable employee demands for wage increases because he deems the stock price to be his most important responsibility, and then leaves the company with over $100 million in bonuses?

As we have seen, truth does not appear to come from on-high. It seems to originate closer to the bottom, from wherever the ever-curious IMPACTS reside. So I would recommend the same attitude also advised by many Ionian Greeks: be very skeptical of proclamations of any kind from anyone. Be very skeptical of what I have to say as well. Study it carefully and look for cracks in the arguments. Hold my feet to the fire.

To change the world, the power of the purse must be utilized. Be selective in your purchases. Do not buy products from companies that display extreme SN tendencies instead of IMPACTS tendencies. Case in point: Wal-Mart. Wal-Mart has been accused of employing questionable business practices in almost every segment of its operations, from buying from sweatshops, to underpaying employees, to harassing and blackmailing vendors, to intimidating union organizing efforts, and to placing unnecessary barriers in the way of employees attaining health insurance. It does not appear to be an IMPACTS-friendly company in any way.

Personally, I wouldn't buy anything from Wal-Mart, even if they were 'giving it away'. Wal-Mart's owners are worth over $30 billion. To me, that is not a success story but rather a story of failure. It appears to be a symbol of the worst kind of hierarchal excesses, a pure accretion of human energy with the barest minimum given in return. If Wal-Mart wanted to offer this book for sale in its stores, I would not allow it, no matter the potential revenue.

Invest only in companies that practice IMPACTS principles in all categories: gender equality in hiring and advancement, environmentally-friendly policies, emphasis on employee well-being and the actualizing of employee potential, and a strong customer-first policy rather than an 'enrich the executives' policy. The purpose of business is to solve problems—to deliver needed and valuable goods, services, and information. It is not to make little-talented, egocentric executives wealthy. So reward the companies that practice good IMPACTS business principles. Also, when possible patronize locally-owned small businesses in your area that follow the same guidelines. Business does not have to be a four-letter word. There will always be great businesspeople in any community who believe in doing things the right way.

Travel to and invest in countries that practice justice-oriented IMPACTS principles or are moving strongly in that direction. Do not enable countries that do not practice human rights, and do not succumb to the pressure to do so. Do not support countries which clearly have hostile intent and behavior towards others, either domestically or abroad. No, we do not have anything approaching a perfect country anywhere in the world, but it should be fairly simple to separate the aggressors from the defenders. Look for accretion. That and aggression usually go hand-in-hand.

IMPACTS have been marginalized politically to a large degree in the U.S. by the current setup of two parties, each on opposite ends of the spectrum. The theory is that occasionally something good may come out of the squabbling. But there is precious little evidence of that. The most prevalent bonds in the U.S. government are between special interests and legislators, which are not the IMPACTS bonds that are needed. IMPACTS bonds are inclusive—SN bonding is exclusive. We need a process that involves the IMPACTS, one that responds to their pressures for change. Where the IMPACTS are involved—science, cutting-edge business and technology, medicine, art and culture—things happen, quickly. We need that pressure in politics too.

The world cannot be the way it was on the island of Crete with the Minoan culture and its IMPACTS society over 3,500 years ago: technological wonders, harmonious lifestyle, beautiful artwork, no defenses, and only a titular king. "The cat is out of the bag"—or I should say, "The SN monster is out of the bag." It is a ferocious force, just as a black hole is ferocious. And it cannot be put back into the bag. Rather, it must be transformed through natural selection with pressure from the IMPACTS.

The 'strong nuclear force' (SN) took control of human societies about 6,000 years ago, and it has not let up since. You will note that agriculture emerged about 10,000 years ago, so it took several thousand years for the SN to develop into a well-defined force. Likewise, if the IMPACTS are to change the trajectory of human behavior, it will take time.

We have allowed the SN to define and control our world, but we really had little choice. As we have seen, that is the natural order of things, unless and until IMPACTS energy steps in to assert itself, thereby creating a different configuration. That is what occurs in the atom when the valence electron bonds with the valence electron of another atom—a new structure, a molecule, is created. And without these molecules, there would be no life as we know it. So bonding is the only way to move toward some kind of permanence in the human sphere that can work against dispersal.

This is what IMPACTS have to do continually—they have to keep connecting with one another and devising peaceful strategies to counter the effects of the SN. For change to occur, there must be an awareness of the situation and its dynamics, identification of critical needs, and action taken to find solutions. It happens on the atomic level—it can and must happen in the human world. Awareness and knowledge are the first steps; then the process can start moving. It will not be fast but there is no alternative because the present configuration is leading us to destruction. A worldwide, grassroots approach is needed. The power of the SN needs to be mitigated and equalized.

IMPACTS Bonding is the Only Way to Neutralize the SN

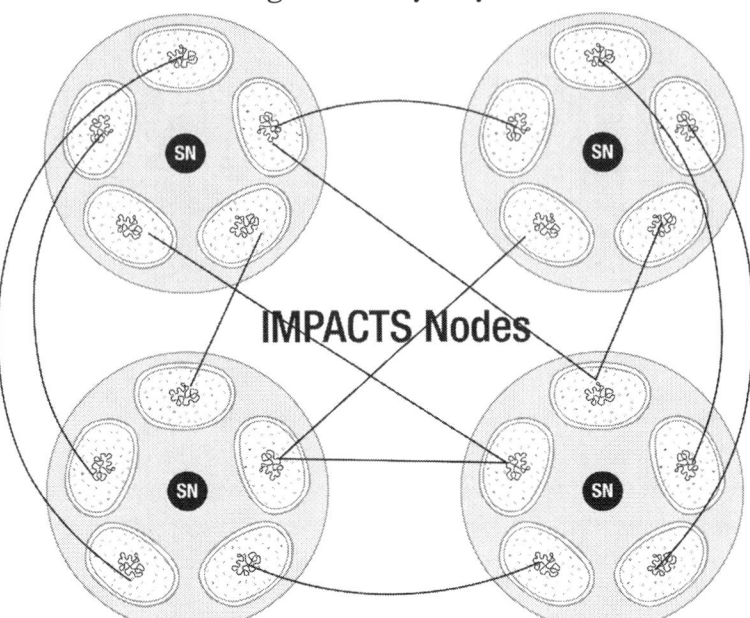

IMPACTS Have to Bypass the SN and Bond with Other IMPACTS Around the World Just as Valence Electrons Bypass the Nucleus to Form Molecules

The present condition of the world, both environmentally and socially, is compelling evidence that the influence of the IMPACTS is not nearly as strong as it has to be. In other words, the bonds that hinder dispersal need to be much more powerful. The environmental and human destruction is SN-induced dispersal, and that can only be counteracted by concentrated and extensive IMPACTS bonding. The SN will not stop until it is stopped, and IMPACTS bonding is the only way to achieve that.

It took several hundred million years for forces to break down the hydrogen atoms that formed 380,000 years after the Big Bang. When the electrons and protons were separated and the universe was reionized, the dark ages of the universe began to come to an end, and light started to shine through. We do not have several hundred million years, but now we do have knowledge and awareness. That is a powerful combination.

IMPACTS can change the world as no other group can. Chloroplasts and mitochondria are remnants from the distant past, but without them, this world would be devoid of most of the life that we see around us. But they are captured, and therefore, the wheels keep turning. The IMPACTS are captured too, but we have a choice in how we respond. We can change the way we support the SN to pressure it to conform more to IMPACTS ideals instead of the ideals of the unfeeling SN.

Whenever you see excellence at any point in the development of modern humans—Stonehenge, the Parthenon, the moon landing, a modern computer, a piece of art—those are the works of IMPACTS. That excellence has served the SN well over the past few thousand years. But it has been mishandled. It is time for the IMPACTS to make sure that their excellence is being used in a way that optimizes the situation—and that situation is life on this planet. If we leave it to the SN, we already have vivid examples of where it will lead.

When the San-shaman was in charge of human affairs, kindness to others meant kindness to all—people, the earth, plants, animals. Justice meant justice for all. Love meant love for all. But the SN wants us to be selective. Justice for the SN means justice as the SN defines it. The SN may want you to see injustice in your immediate environment, but it does not want your eyes to gaze much beyond that. In state-to-state relations, it will define what is just and what is not. I would recommend that you take this attitude: if you are for justice anywhere, you are for justice everywhere.

The energy and ingenuity of the IMPACTS cannot go simply to fixing the 'messes' that the SN causes, or to creating products and services that enable the SN to continue to exert its power and control, often with catastrophic results. It needs to be considered on the front end—what kind of effect will this creative-formative-production have on the Big Picture? The priorities of the first modern humans, the San, must be our priorities: the health of all people and communities, respect for all individuals and their views, sharing and cooperation, respect for and care of nature, and the pursuit of excellence. Those have to be guiding principles, integral to our world, if the IMPACTS are to counterbalance the SN.

Do the IMPACTS have what it takes to bring about more balance, or will a new human species be required? That is not a rhetorical question; I am serious. This species could go the way of previous human species—extinction. Just because we have made it this far does not mean we will continue. We have only been here 130,000 years or so (as homo sapiens sapiens). Some human species lived for one million years or longer, doing the same things day after day. But their effect on the environment was minimal. Ours is prodigious.

If a new human species appeared, it would be very much like our own but with one major difference—the IMPACTS of the next species would be more assertive and less subservient to the SN. It is OK to be a mealybug when the herdsman ant is in control because the ant understands the situation. It does not work when the IMPACTS are mealybugs because the human SN does not understand or have an appreciation for the dynamics involved, though they are exactly the same in both situations.

If the IMPACTS were more assertive and less subservient, then we would have more balance, and therefore a better chance of living in harmony with each other and with the earth. If we continue on the present path, will the forces realign themselves such that the IMPACTS become more prominent? Possibly, but I doubt it.

During the long reign of the San, the tribe controlled the attitudes of the group. No show of self-importance or aggression was permitted. Men and women were equal. Nature was sacred. Animals were revered. Ancestors were worshipped. Now the SN is controlling the attitudes, and what we have does not compare favorably with the San. Men and women are not equal; the display and feeling of self-importance is ubiquitous; human beings are aggressive beyond comprehension; nature is being horribly abused and has been for some time; animals are treated inhumanely and slaughtered recklessly in horrific numbers; and we have lost touch with our San foundational principles. Human bonding is continually stretched thinner and thinner.

When you look at a spiral galaxy such as our Milky Way, you can see that the structure is set. The super-massive black hole is in place in the center, and though star formation is proceeding throughout the galaxy, no appreciable change is seen. But we are not trying to change the shape of the human galaxy or restore it to its original prokaryotic aspects, even if we could. Human civilization has gone well beyond that point, just as the eukaryotic cell has gone beyond the one-celled prokaryotes. What we are looking for is more balance between the SN forces and the IMPACTS forces.

In *Animate Earth—Science, Intuition, and Gaia*, Stephan Harding says that hydrogen is so light with its one proton and one electron, if it were not captured in molecules, it would just float off into the cosmos. And as we know, if we did not have hydrogen, we would not have anything. Likewise for its electron—if it were not captured by the proton, it would probably fly away also, to reunite with other electrons. It is the same with the IMPACTS. Throughout human development, they have been mostly residing on the periphery. If they were not captured by the SN, they would still be out there. Many of them still are.

This world is often not easy for IMPACTS. They see problems that others do not see, problems that distress them and problems that they are often prevented from resolving, even though they have the skills. The world can be constricting when what IMPACTS need is freedom, freedom to innovate and produce, to help others and humankind, to trade, to learn and to teach, to express themselves and their opinions, to create and enjoy art and culture, to avoid wars and conflicts, to live independently in a community environment, to roam or travel, and freedom from hierarchal dictates or an overbearing structural-nucleus. IMPACTS need room to breathe just as any plant needs room to grow. But just as plants can grow in a crack in the road, so too can IMPACTS survive the most extreme conditions and still contribute to humankind. They are tenacious, resilient, and resourceful.

In Greek mythology, Sisyphus was a king cursed and punished by being forced to roll a huge boulder up a hill, only to watch it roll back down again, and to repeat this throughout eternity. Are IMPACTS consigned to the role of Sisyphus for eternity, or can they unshackle themselves and get out of the structure enough to save it? If the electron of the hydrogen atom had not stepped 'outside the lines' and connected with the electron of another hydrogen atom, we would have no molecules, no stars, and no life. The universe would still be dark. IMPACTS have to do the same—they have to step outside the lines and continue connecting with other IMPACTS, forming powerful, concentrated, anti-dispersal bonds.

After all is said and done, what do we really want? We want people who can see clearly what is needed, and who will then take the necessary steps to fix the problem, even if it means stepping outside the lines. We want people with backbone who are not afraid to stand for principles, people like Martin Luther King, Jr., and Gandhi, and Nelson Mandela. We want people who work hard to hinder SN-induced dispersal. We want 2nd Law-breakers.

Now is the most critical time in the history of the world, and IMPACTS will have to find unique ways to use their innovation and bonding to help build a more sustainable future for human beings. It is up to the IMPACTS just as it has always been.

Appendix

A Review of the San Tribe and the San-shaman

The Progenitors of Today's IMPACTS

General Attitude of San Tribe Members
To help maintain health and well-being of individuals and the tribe through strong, harmonious bonding with each other, the shamans, nature, deceased ancestors, and spirits.

San Tribe General Traits
- Hands-on
- Nurturing
- Competent
- Compassionate
- Multi-tasking
- Optimistic
- Positive
- Friendly
- Humble
- Semi-autonomous to autonomous
- Androgynous
- Hard-working
- Dependable-trustworthy
- Frugal
- Egalitarian
- Ethical
- Protective of each other
- Conscientious
- Empathy and sympathy for others
- For the underdog
- Mentally, emotionally, physically tough
- Sharing—benevolent—altruistic

San-shaman Main Task (role)
To promote and facilitate well-being and health for individuals and the tribe through the maintenance of strong, harmonious bonding among people and with nature, deceased ancestors, and spirits.

Accomplished by: Solving problems, sometimes through a trance journey, and delivering the solutions in an expeditious yet caring manner.

San-shaman Traits
- Sense of urgency
- Almost selfless
- Focused on the here and now but mindful of the past and future
- Alert—aware of needs in the environment
- Contextual intelligence—awareness of possibilities in the environment
- Problem-solver
- Leadership qualities
- Takes initiative
- Passionate about helping others—highly motivated
- Cautiously optimistic
- Positive
- Innovative-creative-artistic
- Tireless
- Striving for objectivity
- Open to new ideas
- On a mission
- Good listening skills
- Dedicated-committed
- Detail-oriented (precise)
- Strong conscience
- Efficient
- Flexible
- Tenacious
- Unassuming-unpretentious
- Eager to inform others of solutions to problems
- Protective of tribe
- Takes charge—quietly aggressive
- Priorities in order
- Regular person—no special treatment from tribe
- Direct—no-nonsense
- People-person

Special Capabilities Needed by San-shaman
- Planning
- Organizational
- Diagnostic – what are the problems?
- Analytical - what needs to be done to solve them?
- Communicational – often nonverbal such as art and dance
- Scientific attitude – trial and error
- Research
- Setting up systems to duplicate efforts and solutions

Philosophy and Goals
- Alleviation of suffering
- Answering critical needs and facilitating solutions
- Taking from Unwell to Well or Unusable to Usable
- Maintenance of health and well-being for others and the tribe
- Helping the person and tribe realize the greatest potential
- Service to others

Lifestyle
- Nomadic with logistical concerns
- Close-knit community
- Lived in circular environments
- Traveled and lived near fresh water
- Obvious close association with nature and animals

Strong priorities
- Relationships with extended family and other tribal members
- All children
- Animals for food and power
- Cooperation with other tribes to aid well-being for all
- Journeys to the spirit world for answers

Core of the Shaman's Efforts
- Facilitating a transformation or transformations

Overall attitude
- Not satisfied with the status quo—passionately working for positive change and continual improvement in people, processes, results, and the total environment.

Glossary

2nd Law of Thermodynamics

The 2nd Law is the strongest known force or tendency in the universe. Basically the law states that concentrated energy will disperse if not hindered from doing so. An example is a tire filled with pressurized air. If the tire is punctured, the pressurized air immediately escapes until there is equilibrium between the inside of the tire and the outside of the tire. That is how the 2nd Law works—the tendency is toward homogeneity.

The 2nd Law also has an antithesis which is dispersal-hindering, or else we wouldn't be here. This appears to be accomplished on earth mostly through the bonding of valence electrons, resulting in molecules and compounds. What we see everywhere we look is violating the 2nd Law. So anti-dispersal is powerful as well but ultimately it will be defeated—or so it is believed.

San Tribe

The oldest modern human group on earth at over 100,000 years old. The San developed in Africa and lived a close-knit, community-oriented, nomadic life with the San-shaman at the center.

The San-shaman

Life for the San revolved around the shaman. There were male and female shamans but usually more males. The shaman's first priority was healing and that led to immense innovation. Travel to the spirit world through trance was frequently utilized in the quest for solutions. The shamans were the main 'dispersal-hinderers'.

IMPACTS

People who have the same basic philosophy and attitude toward life as did the San tribe and the San-shaman. Some IMPACTS will be more like the problem-solving shaman and others will be more like tribal members who helped deliver solutions. The entire profile is based upon helping people get well and stay well. The IMPACTS are today's 'dispersal-hinderers'. They hold human society together all over the world, or certainly attempt to do so.

IMPACTS is plural. The singular of IMPACTS is IMPACTS-person. She is not an IMPACT—she is an IMPACTS-person or IMPACTS-individual.

Creative-formative-productive energy (CFPE)

The energy that creates or enables the creation of a structure—any structure. It is mostly female energy and I believe it starts with the electron. It is also the energy that innovates, maintains, improves, and changes the structure. It has a strong protective element as well. Think of the traits of human mothers. That is CFPE and IMPACTS energy.

IMPACTS Energy

The same as CFPE. It appears to be the principal antidote to the 2nd Law and dispersal tendencies. It is also 'new beginnings' energy.

IMPACTS Nodes

Pockets of IMPACTS energy. Valence electron bonding can be thought of as an IMPACTS node as can a family and an entrepreneurial group. An IMPACTS node is composed of a strong IMPACTS energy component, and bonding of these nodes is an enhancement of IMPACTS energy.

IMPACTS Dynamic

The dynamic in the universe that continually and constantly seeks to balance and optimize the situation. This usually occurs from a peripheral location as it relates to the SN, or structural-nucleus, of the structure. It is mostly a female dynamic. The valence electron is a manifestation of the IMPACTS Dynamic. So are the IMPACTS in human society though they are certainly not as predictable as is a valence electron.

SN

The structural-nucleus of an entity. The entity can be a country, a company, an industry, an organization, a school system, or anything else. The SN will be the power-control center of the entity. It is predominantly male energy like the proton (nucleus) of the hydrogen atom. The SN can be singular or plural.

Extreme SN

An SN with extreme power-control aspects, usually accompanied by violent tendencies.

SN-IMPACTS

Those IMPACTS who are philosophically closer to the structural-nucleus or SN of society, which is the power-control center. In many ways, they are captured in service to the SN.

P-IMPACTS

Those IMPACTS who reside more on the periphery of society. Sometimes this will be of a physical nature, but always of a philosophical nature. P-IMPACTS are less affected by the SN of society than are SN-IMPACTS.

Accretion

Accretion is a ubiquitous phenomenon in the universe in which creative-formative-productive energy (CFPE) is taken by an accretive entity (accretor) and used for its survival. An example is a binary star system where the dying star is accreting matter and energy from the healthier star. It is also part of the major dynamic that exist in human society. There are accretors and there are producers. The IMPACTS are usually on the producing side though SN-IMPACTS can be producers and accretors. The SN is predominantly accretive.

Nucleated Structure

A structure that possesses a nucleus or structural-nucleus which controls the structure. The nucleus depends on IMPACTS energy or CFPE elements to supply the structure with operational energy. An example is a eukaryotic cell which depends on mitochondria to power the cell. Mitochondria are derived from prokaryotic bacteria which have no nucleus. A nucleated structure appears to be the result of the interaction of the 2nd Law of Thermodynamics (dispersal force) and the dispersal-hindering force, which is IMPACTS energy, or CFPE.

Pre-nucleated Structure

A structure that functions without a nucleus or with a loosely-defined nucleus, such as a prokaryotic cell. The structure powers itself with IMPACTS energy or CFPE and handles the control functions, such as they are. The San tribe is an example of a pre-nucleated or prokaryotic structure. When the structure attains a certain level of IMPACTS energy and male energy is nearby, male energy usually captures the female IMPACTS energy, forming a nucleated structure. But if male energy is not sufficiently present or is not interested, the structure may remain prokaryotic.

Selected Bibliography

Inside the Neolithic Mind, David Lewis-Williams and David Pearce,
Thames and Hudson, 2005

The Mind in the Cave, David Lewis-Williams,
Thames and Hudson, 2002

How Art Made the World, Nigel Spivey,
Basic Books, A Member of the Perseus Books Group, 2005

Guns, Germs, and Steel, Jared Diamond,
W. W. Norton and Company, 1999

A Short History of Nearly Everything, Bill Bryson,
Broadway Books, A Division of Random House, Inc., 2005

Animate Earth – Science, Intuition, and Gaia, Stephan Harding,
A Sciencewriters Book, Chelsea Green Publishing Company, 2006

Full House: The Spread of Excellence from Plato to Darwin,
 Stephen Gould, Harmony Books, 1996

The Knowledge Web, James Burke,
A Touchstone Book, Simon and Schuster, 1999

The Day the Universe Changed, James Burke,
Back Bay Books, Little, Brown and Company, 1995

Good to Great, Jim Collins,
HarperBusiness, 2001

Built to Last, Jim Collins and Jerry I. Porras,
HarperBusiness Essentials, 2002

The Millionaire Mind, Thomas J. Stanley, Ph.D.,
Andrews McMeel Publishing, 2001

Please Understand Me II, David Keirsey,
Prometheus Nemesis Book Company, 1998

DVDs

Japan's Mysterious Pyramids, History Channel, 2000

Marco Polo - Journey to the East, Biography, A&E, 1995

Digging for the Truth—Giants of Easter Island, A&E, 2005

Cities of the Underworld—Beneath Vesuvius, A&E, 2007

In Search of History—Ancient Invention, A&E, 1997

Cities of the Underground—Secret Pagan Underworld, A&E, 2007

Vikings—Fury from the North, A&E, 2000

Enduring Mystery of Stonehenge, A&E, 1998

Ancient Discoveries—Machines III, A&E, 2006

Ancient Egypt—Modern Medicine, A&E, 2006

Dreamtime of the Aborigine, A&E, 1997

Ants: Little Creatures Who Run the World, NOVA, WGBH, 2007

Guns, Germs, and Steel, Jared Diamond, National Geographic, 2005

How Art Made the World, Nigel Spivey, BBC Video, 2006

African American Lives, PBS Home Video, 2006

The Universe – Complete Season One, The History Channel, 2007

How The Earth Was Made, The History Channel, 2007

Living with Wolves, The Discovery Channel, 2005

Super Volcano, The Discovery Channel, 2003

In Search of Eden, The Discovery Channel, 2002

The Medici – Godfathers of the Renaissance, PBS Home Video, 2003

The Greeks, PBS Home Video, 2005

Da Vinci & Mysteries of the Renaissance, Questar

If We Had No Moon, The Discovery Channel, 2003

The Real Eve, The Discovery Channel, 2002

Walking With Cavemen, The Discovery Channel, 2003

Stonehenge and the Ancient Britons, Kultur

Origins – Fourteen Billion Years of Cosmic Evolution, NOVA, Hosted by Neil DeGrasse Tyson, 2004

Einstein Revealed, NOVA, 2004

Rome – Power and Glory, Questar

Ice World, The Discovery Channel, 2003

Hyperspace, BBC Video, 2001

TV series—The following were available from iTunes via the History Channel website at various times from 2007-2009:

Underground Apocalypse: Cities of the Underworld	Season 2
Viking Underground: Cities of the Underworld	Season 2
Prophecies from Below: Cities of the Underworld	Season 2
Maya Underground: Cities of the Underworld	Season 2
Mysteries of the Moon: The Universe	Season 2
The Milky Way: The Universe	Season 2
Dark Matter: The Universe	Season 2
Unexplained Mysteries: The Universe	Season 2
Wildest Weather In the Cosmos: The Universe	Season 2
Biggest Things in Space: The Universe	Season 2
Gravity: The Universe	Season 2
Cosmic Apocalypse: The Universe	Season 2

Websites—Thousands but here are a few recent ones:

http://www.chemistry.mcmaster.ca/esam/intro.html
http://www.esa.int/easCP/SEMKTX2MDAF_index_2.html
http://www.nytimes.com/2008/05/22/science/22nova.html?bl=&ei=
&adxnnl=1&adxnnlx=1212931509-+8eI1vF/XMq2epXb6PhOSg
http://www.sciencedaily.com/releases/2007/04/070402103058.htm
http://www.nrc-
cnrc.gc.ca/highlights/2005/0501electronimaging_e.html
http://cnx.org/content/m12584/latest/
http://www.space.com/scienceastronomy/060109_event_horizon.html
http://hnn.us/articles/40538.html
http://archive.ncsa.uiuc.edu/Cyberia/NumRel/BlackHoleAnat.html
http://www.aspergers.com/aspclin.htm
http://www.cv.nrao.edu/~abridle/dragnparts.htm
http://searchcio-
midmarket.techtarget.com/sDefinition/0,,sid183_gci341428,00.html
http://www.sciam.com/article.cfm?id=matter-antimatter-split-hi&pri
http://www.nsf.gov/od/lpa/news/02/pr0288.htm
http://science.nasa.gov/newhome/headlines/ast07sep99_1.htm
http://mr.caltech.edu/media/Press_Releases/PR11956.html
http://www.pslc.ws/radical.htm
http://physicsworld.com/cws/article/news/19793
http://www.space.com/scienceastronomy/sun_shine_030403.html
http://spaceflightnow.com/news/n0204/12darkmatter/
http://news.nationalgeographic.com/news/2003/06/0619_030619_dar
kside.html
http://www.ph.surrey.ac.uk/astrophysics/files/how_stars_form.html
http://helios.gsfc.nasa.gov/compos.html
http://space.newscientist.com/channel/astronomy/dn13216-milky-
ways-antimatter-linked-to-exotic-black-
holes.html?feedId=astronomy_rss20
http://www.sciam.com/podcast/episode.cfm?id=9FC85F47-E26D-
F464-F9CB72C248931F27
http://www.aanda.org/content/view/304/42/lang,en/
http://www.sciencedaily.com/releases/2008/04/080424130707.htm
http://www.sciencedaily.com/releases/2007/06/070606113357.htm
http://physicsworld.com/cws/article/news/26452
http://www.nrao.edu/pr/2003/j1148/index-p.shtml
http://www.astronomical.org/astbook/binary.html
http://csep10.phys.utk.edu/astr162/lect/blackhole/blackhole.html
http://www.astronomynotes.com/galaxy/s14.htm
http://altreligion.about.com/library/weekly/aa041506a.htm

Made in the USA
Charleston, SC
01 February 2011